DOE/EM-0362

Accelerating Cleanup

Paths to Closure

U.S. Department of Energy
Office of Environmental Management

June 1998

Printed with soy ink on recycled paper

The Department of Energy is responsible for the world's largest environmental cleanup program. This enormous technical challenge is a national priority that must be based on thorough, scientific analyses. Our nation must find solutions backed fully by Congress, state and local governments, Tribal Nations, regulators and stakeholders.

In the last five years, the Department has made substantial progress in systematically defining the scope, schedules, and life-cycle costs to meet this challenge and creating a step-by-step work plan to tackle it. *Accelerating Cleanup: Paths to Closure* outlines the Energy Department's evolving and dynamic cleanup program based on site-developed, project-by-project forecasts of the scope, schedule, and costs to complete the 353 projects that currently define the cleanup program.

Meeting the enormous cleanup challenge requires an enduring national commitment. This program will not succeed unless sufficient and consistent resources are available over the long term and unless the Department continually seeks efficient and cost-effective ways of doing business. As this report demonstrates, a long-term budget at current levels would enable the Department to accelerate the cleanup and closure of many of its sites. The accelerated cleanup would reduce health and environmental risks, make sites available for community re-use, and maintain our compliance to federal and state laws and agreements. Securing sufficient resources and achieving these commitments will remain a challenge. If adequate resources are not sustained, progress will be slower, health and environmental risks will last longer, and cleanup ultimately will cost even more.

Consistent with this Administration's initiative to work smarter, the Department's Environmental Management program will continue to seek opportunities to complete our cleanup work as quickly and efficiently as possible. At the same time, we will not accelerate cleanup by compromising cleanup standards or the health and safety of our workers.

The Department values our partnership with stakeholders, regulators, the Congress and Tribal Nations in developing and implementing our cleanup program at each site. The Department will continue to seek advice, support and guidance from our partners as the *Paths to Closure* report is updated.

Federico Peña
Secretary of Energy

Table of Contents

Executive Summary

Vision

By 2006, the Environmental Management program intends to complete cleanup at most of its 53 remaining sites. At the 10 remaining sites, including our five largest sites, treatment will continue for the remaining "legacy" waste streams. This vision will drive budget decisions, the sequencing of projects, and the actions needed to meet program objectives. This vision will be implemented in collaboration with stakeholders, regulators, and Tribal Nations.

The Challenge

Cleanup of the radioactive, chemical, and other hazardous waste left after 50 years of U.S. production of nuclear weapons is the largest environmental management program in the world. Only in the last five years has the Department of Energy (DOE) made substantial progress in systematically defining the technical scope, schedules, and life-cycle costs of meeting this challenge, and creating a step-by-step work plan to tackle it.

The Department of Energy, its stakeholders, its regulators, Tribal Nations, the Congress, and the American people want to accelerate and finish the job of cleaning up DOE's sites. At the same time, we all continue to share the goal of placing the safety of our workers, our communities, and the environment first among all other priorities.

Accelerating Cleanup: Paths to Closure (hereinafter referred to as *Paths to Closure*) provides, for the first time, a site-by-site, project-by-project projection of the technical scope, cost, and schedule required to complete all 353 projects at DOE's 53 remaining cleanup sites in the United States. These projections are essential for better management—they provide critical information on technical activities, budgets, worker health and safety, and risk to inform regulators, state and local officials, stakeholders, Tribal Nations, and others. Like DOE itself, all these groups need an understanding of the technical requirements for meeting DOE's obligations and agreements. We can then work together to clean up as many sites as possible, as quickly and safely as possible. Our goal is to clean up more than 90 percent of our sites by 2006. It is important to note that the "closure" of a site does not end DOE's responsibility. In most cases, DOE will continue long-term surveillance and monitoring activities to ensure that human health and the environment are protected.

Resources are limited. Technical risks are often high, and schedules for meeting compliance agreements are often very ambitious. For the first time, we—DOE officials, stakeholders, regulators, Tribal Nations, and the Congress—have a comprehensive management tool that can inform us of the consequences of our choices. *Paths to Closure* provides:

- An integrated path forward for the management of DOE's Environmental Management (EM) program[1], based on a site-by-site, project-by-project, life-cycle foundation;
- A basis to evaluate EM's annual budgets in the context of long-term cleanup and closure requirements and projections;
- A response to Congressional requests for a supportable management strategy on the EM program; and
- A response to the concerns of stakeholders, regulators, and Tribal Nations.

Paths to Closure reflects the most recent evolution of DOE's ability to accurately project the cost, schedule and scope of its massive cleanup effort. *Paths to Closure* is part of a continuum from the first life-cycle cost estimates and risk analyses contained in the *Baseline Environmental Management Report* (BEMR) that initiated the first national dialogue on these issues. *Paths to Closure* is a critical management tool that reflects project-by-project work plans of each of 353 projects at DOE cleanup sites nationwide. Current life-cycle estimates for cleanup, based on the assumptions described in this report, total $147 billion.

Paths to Closure also reflects DOE's strengthened and more organized commitment to listen and respond to stakeholder, regulator, Tribal Nation, and internal DOE concerns. The result is a more realistic projection of where we are headed, how we can accelerate cleanup and closure, and what the technical, policy, and other barriers are to the further acceleration of those goals. This report incorporates comments and guidance received from stakeholders, regulators, and Tribal Nations on the draft circulated in February 1998.

A key change to the February draft is the addition of a discussion on the Environmental Management program's decision-making process and *Paths to Closure*'s relationship to that process. This report also includes a new chapter summarizing comments received on the draft and describing changes made to the draft. The basic work scope, cost, and schedule data supporting this report are the same as those used to develop the February draft *Paths to Closure*.

Chapter 1 describes in more detail the process by which *Paths to Closure* has been developed and what it hopes to accomplish, its relationship to the Environmental Management decision-making process, and a general background of the Environmental Management mission and program. Chapter 2, "Baseline Scope, Schedule, and Cost," describes how the site-by-site projections were constructed, and summarizes, for each of DOE's 11 Operations/Field Offices, the projected costs and schedules for completing the cleanup mission. Chapter 3 presents summaries of the detailed cleanup projections from three of the 11 Operations/Field Offices: Rocky Flats (Colorado), Richland (Washington), and Savannah River (South Carolina). The remaining eight Operations/Field Office

[1]Throughout this document, the phrase "Environmental Management program" or "EM program" refers to operations at both the Headquarters and site level. Section 1.3 explains the relationships of Headquarters and site levels in the EM program.

summaries are in Appendix E. These summaries are built on the projections for the individual projects and sites that these offices oversee.

Chapter 4, "Meeting Programmatic Challenges," reviews the cost drivers, budgetary constraints, and "performance enhancements" underlying the detailed analysis of the 353 projects that comprise EM's accelerated cleanup and closure effort. Chapter 5 describes "A Management System To Support the EM Program." Chapter 6 provides responses to the general comments received on the February draft of this document. Specific comments will be addressed in letters to the organizations providing the comment.

Relationship of *Paths to Closure* to the EM Decision-making Process

Public comments on the 1997 *Accelerating Cleanup: Focus on 2006 Discussion Draft* (hereinafter referred to as the *Discussion Draft*) and the February 1998 draft *Paths to Closure* report requested clarification on the decision-making process for the work described in *Paths to Closure*. Decisions in the EM program are driven by various statutory mandates, most notably the Comprehensive Environmental Response, Compensation, and Liability Act (CERCLA) and the Resource Conservation and Recovery Act (RCRA). Most decisions are made at the site level (with appropriate Headquarters oversight). Other decisions are made at the Headquarters level because of their complex-wide implications. In many cases, ultimate decision-making authority, in the sense of final approval authority, resides with U.S. Environmental Protection Agency (EPA) or state regulators.

Public participation is an important element of the EM program's decision-making process. The National Environmental Policy Act (NEPA) requires federal agencies to consider the environmental impacts of their proposed actions. NEPA also requires that the public be informed of, and have an opportunity to comment on, major federal actions significantly affecting the environment. Consistent with its obligations under NEPA, the EM program performs an appropriate level of environmental review in connection with its projects, with opportunities for public involvement. For projects managed under CERCLA, EM relies on the CERCLA process to incorporate NEPA values.

Paths to Closure outlines EM's current estimate of the scope, schedule, and costs for each site to complete the cleanup program. The estimate includes projects for which key decisions have been made pursuant to CERCLA, RCRA, or other statutes, and projects where such decisions have yet to be made. Where decisions have not yet been made, sites make **assumptions** (e.g., site planning end states) about how those cleanup actions might be carried out so that sites can define work and develop schedule and cost estimates. In those cases where decisions have not yet been made, the Environmental Management program will follow the decision-making processes called for by the relevant statutory authority that governs the activity in question (e.g., CERCLA or RCRA) with appropriate environmental review.

Paths to Closure also includes cost estimates for federal salaries, investments in science and technology, and miscellaneous support functions. EM sites and EM Headquarters make decisions through the budgetary process on the scope and pace of work for these activities. Stakeholders and Tribal Nations will have significant opportunities to participate in all decision-making processes.

Projected Scope, Schedule, and Cost

Paths to Closure contains the Environmental Management program's detailed projections on the scope, schedules, and costs at each site for the cleanup of contaminated soil, groundwater, and facilities; treating, storing, and disposing of waste; and effectively managing nuclear materials and spent nuclear fuel. These projections account for, where possible, future decisions that must be made and define the degree of technical and scope uncertainties.

A key component of *Paths to Closure* is the development of projections—or "baselines" (as estimates of individual projects are called). The projections include descriptions of the work to be accomplished, schedules (including interim milestones), and cost estimates for each project. Chapter 2 of this report provides summary information on the scope, schedule, and cost of the Environmental Management program, as derived from these baselines. The division of all cleanup work into projects and the establishment of formal projections, or baselines, represents a significant shift in DOE's approach to environmental management. The process of establishing specific projects and baselines with defined scope, schedule, and cost projections has resulted in significant reductions in EM life-cycle cost estimates.

Developing cost, schedule, and scope projections also requires identifying either an actual or, more often, a planning-based cleanup "end state" for each site. The cleanup of a site is considered to be complete—to have reached its end state—when it has been cleaned up in accordance with agreed-upon cleanup standards. (Additional elements of this definition are provided in Chapter 1.) To develop a cost, schedule, and scope projection for a project, some assumptions have been made about the desired end state. The projections made for this document are based not only on end states that are consistent with existing agreements and applicable regulations but also on planned end states based on assumptions for the many sites still in the process of working with stakeholders, regulators, and Tribal Nations to finalize agreed-upon end states. Many end states will change for a number of reasons, including the development of new technologies, more economical cleanup approaches, and changes in the interests of stakeholders, regulators, and Tribal Nations.

For the first time, every site has a critical closure path, identifying the key technical and programmatic activities that must occur before closing a site. Also for the first time, each site has waste and materials disposition maps that describe each waste stream, the steps for processing or managing the wastes, and where the wastes are intended to be permanently disposed (if known). And finally, for the first time, DOE has identified the potential roadblocks on the critical closure path, by identifying technological uncertainty and the degree of intersite dependence, among other factors.

Projections of scope, schedule, and cost contain the data necessary to establish an estimated life-cycle cleanup cost and a completion date for EM work at each site. *Paths to Closure* provides a funding guideline of $5.75 billion per year for the entire EM program, starting in FY 1999. Site funding needs in excess of the guideline vary from year to year, as is shown in Exhibit 4-2 of this document. No increases are included for future inflation, so in "real" terms (i.e., in terms of constant FY 1998 dollars), the amount of funding decreases every year.

With this funding guideline, the sum of the life-cycle cost estimates for the current 353 projects is about $147 billion between 1997 and 2070. Of this amount, about $57 billion would be expended through 2006; about $90 billion would be expended from 2007-2070. The table below provides a summary of these costs, by Operations/Field Office and time frame.

EM Costs by Operations/Field Office

Operations/ Field Office	Estimated EM Costs (1997-2006)	Estimated EM Costs (2007-2070)	Total Estimated EM Costs (1997-2070)	Number of Sites Completed	
	(All costs in billions of constant 1998 dollars)			1998-2006	After 2006
Albuquerque	2.1	2.0	4.1	12	1
Carlsbad	1.8	5.9	7.7	0	1
Chicago	0.3	0.0	0.3	5	0
Headquarters/ National Programs	5.7	5.6	11.3	NA	NA
Idaho	5.0	11.3	16.3	0	1
Nevada	0.9	1.3	2.2	8	2
Oakland	0.7	0.3	1.0	8	1
Oak Ridge	5.4	7.7	13.1	3	2
Ohio	4.6	0.2	4.8	5	1[a]
Richland	13.0	37.3	50.3	0	1
Rocky Flats	5.3	1.0	6.3	0	1[b]
Savannah River	12.0	17.7	29.7	0	1
TOTAL[c]	57.0	90.3	147.3	41[d]	12
				53	

[a]The one site after 2006 is the Fernald Environmental Management Project (FEMP). It is expected that cleanup at FEMP also will be completed before 2006, although the baseline currently indicates completion in 2008.

[b]The current baseline for the Rocky Flats Environmental Technology Site reflects a 2010 closure. However, the baseline is being revised to reflect the commitment to complete closure by 2006.

[c]Individual costs may not sum to totals due to rounding.

[d]With the accelerated goal of cleaning up the Rocky Flats Environmental Technology Site and the Fernald Environmental Management Project (by 2006 and 2005 respectively), the number of sites completed by 2006 would be 43.

In addition to the $147 billion *Paths to Closure* life-cycle cost estimate, stakeholders have asked for other costs associated with the EM program, but not included in *Paths to Closure*, to be identified. Two examples of these costs are:

- $8.1 billion associated with newly-generated waste generated after FY 2000. *Paths to Closure* was developed under the assumption that EM would transfer these costs back to the generators after FY 2000.

- $8.7 billion associated with deactivation and decommissioning of excess facilities not currently under EM jurisdiction. DOE is considering the transfer of additional surplus facilities to the EM program beginning in FY 2002 with limited exceptions occurring before that date. If and when such transfers occur, EM will develop projects and adjust current assumptions to account for these facilities and to include these costs in future updates to *Paths to Closure*.

Chapter 3 provides more detailed scope, schedule, and cost information for sites under the jurisdiction of three of DOE's Operations/Field Offices. Appendix E provides information on the remaining eight field offices. The more detailed site versions of *Paths to Closure* provide still further details.

Numerous cleanup activities will continue beyond 2006. Projections reveal that at the Hanford Site in Washington, the Idaho National Engineering and Environmental Laboratory, and the Savannah River Site in South Carolina, about half the costs will be incurred after 2006 for treatment and disposal of high-level and transuranic waste. Although some activities will not be completed by 2006, a primary goal of *Paths to Closure* is to reduce outyear costs. At the end of FY 1997, 60 of the 113 contaminated sites had been cleaned up. An additional 43 sites are estimated to be cleaned up between 1998 and 2006—for a total of 103 cleaned up sites by 2006 (see box). Long-term cleanup activities will continue at the remaining 10 sites. Major cleanup goals for 2006 include:

- Remediation of 80 percent of all release sites, that is, specific locations or areas where contaminants may have been released to the environment;

- Stabilization of all nuclear materials and spent nuclear fuel and completion of all preparations for their ultimate disposition; and

- Completion of all cleanup activities at some major sites, for example, the Rocky Flats Environmental Technology Site, the Fernald Environmental Management Project, the Miamisburg Environmental Management Project, and the Weldon Spring Site.

Meeting Programmatic Challenges

To reduce the costs of this massive cleanup effort, the Environmental Management program continues to seek significant opportunities to accelerate cleanup without jeopardizing the safety of workers, communities, or the environment. *Paths to Closure* addresses the need to continuously seek "performance enhancements," i.e., productivity improvements that will allow DOE to accelerate cleanup and closure schedules, and lower overall life-cycle cleanup costs. The EM program is focusing on six specific mechanisms to help achieve additional performance efficiencies (see box).

Performance Enhancement Mechanisms

Mechanism	Achieves Efficiency By...
Technology Deployment	Introducing less expensive and/or more effective cleanup technologies.
Integration	Identifying better ways to transfer and manage wastes among sites.
Project Sequencing	Completing projects with high "upkeep" costs.
Pollution Prevention	Reducing waste volumes and associated disposal costs.
Contract Reform	Creating incentives for contractors to work less expensively.
Lessons Learned	Increasing productivity based on lessons learned.

Accelerating cleanup even further than is projected in *Paths to Closure* will certainly happen, although the degree of acceleration is difficult to predict. For example, DOE and its stakeholders and regulators in Colorado have established an accelerated goal of cleaning up and closing the Rocky Flats Environmental Technology Site by 2006—four years earlier than the current baseline indicates. DOE will attempt to set similar acceleration goals at other cleanup sites. Credible acceleration goals will be based on the likelihood of achieving technology deployment, intersite integration, and other productivity improvements. Chapter 4 of this report discusses enhanced performance mechanisms and goals in greater detail.

Although *Paths to Closure* is not a budget document, it is designed to be an integral part of the annual and multi-year DOE budget development process. The projections prepared for each site are the basis upon which future resource allocation decisions can be made. In building future budgets, differences will emerge between the cost projections established in this and future *Paths to Closure* reports, and budget allocations to DOE from the President and the Congress. *Paths to Closure* gives EM, its stakeholders, regulators, and Tribal Nations, and the Congress the management tools we need to understand the consequences of our choices—the effects on life-cycle costs and closure date schedules of alternative near-term and outyear budget scenarios.

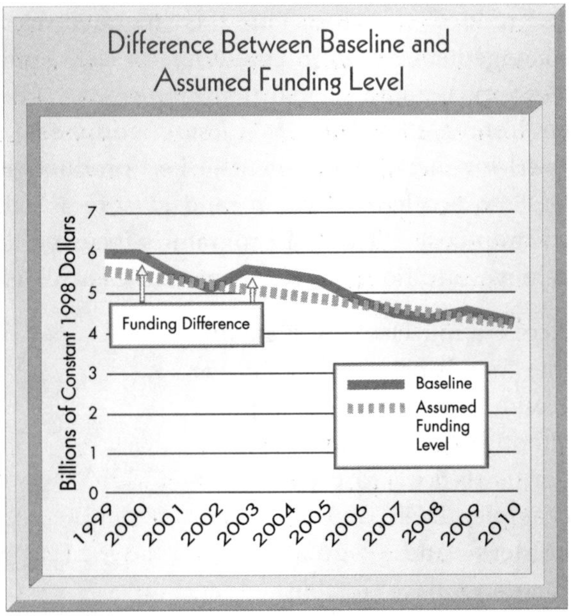

Paths to Closure provides a funding guideline of $5.75 billion per year for the entire EM program, starting in FY 1999. This figure was set in October 1997, *prior to DOE receiving its FY 1999 and outyear budget targets from the President.* It was essential to establish a funding profile at that time in order to produce this report on schedule. In some cases, sites exceeded the $5.75 funding guideline to meet compliance commitments. One critical budget and resource allocation question is how the EM program will make up the difference between the funding guideline of $5.75 billion, and the requirement for more than that in several future years to meet compliance agreements and other commitments. An even more difficult question is what would happen if the funding guideline of $5.75 billion per year were not met. The chart above converts the $5.75 billion per year in "current" (or "nominal") dollars, to "constant" FY 1998 dollars—thus showing how inflation lowers the "real" amount of money available each year. The higher "baseline" level of funding is that which is required based on the projections from each of the 353 projects. The gap between the two is $3.9 billion (in constant FY 1998 dollars) between 1999 and 2006.

The first step in meeting this challenge is the aggressive application of the productivity improvements—the performance enhancements—described above and in Chapter 4. The performance enhancements are expected to include improvements in the efficiency of day-to-day operations, better application of science, the deployment of new technologies, and streamlined approaches—to be developed with regulators—for managing waste and cleaning up contaminated areas.

If performance enhancements are not sufficient to address funding differences at specific sites, and if additional funding were not obtained, EM would pursue several options. In cases where new work is required immediately to protect safety and health and where related costs exceed available appropriations, the Department will shift funds from lower priority activities to ensure that public health and safety are adequately protected.

In future years where larger funding differences are projected, the Department intends to work with the Office of Management and Budget (OMB) and the Congress to seek additional funds for vitally important missions. Also, DOE will propose shifting outyear funding from completed sites to other sites. No matter how successful these efforts are, however, the discipline of working within binding budget ceilings means that the EM program must engage in an active dialogue with stakeholders, regulators, and Tribal Nations about activities and programs at each of the sites—and collectively make hard choices regarding priorities.

A Management System to Support the EM Program

The Environmental Management program is developing a formal integrated management system to more closely align *Paths to Closure* and the annual budget formulation process. This system will allow the Environmental Management program to use a single framework for all activities linked to planning, the budget formulation and execution process, and performance measurement. For the first time, EM is working toward the implementation of a truly integrated life-cycle database containing most of the data the field provides to Headquarters. Chapter 5 of this report describes the EM management system components of the process in greater detail. Some of the new management tools include:

- Waste/Material Disposition Maps (or flow charts), which are conceptual approaches to the environmental remediation of contaminated soil, groundwater, and buildings; and for the storage, treatment, and disposal of all waste and material at all sites;
- Critical Closure Paths, which are the schedules of activities that must be completed on time in order for cleanup to be accomplished;
- Identification of specific science and technology needs, to help reduce the cost and risk of specific projects by developing improved cleanup technologies; and
- Programmatic Risk Assessments, which provide a measure of the risks associated with accomplishing the work and meeting schedules and cost estimates.

As the cleanup program moves forward, the quality of the data on which the above tools are based continues to improve. *Paths to Closure* represents a significant refinement over the national *Discussion Draft* and the site *Discussion Drafts* published in June 1997. Project baselines, the heart of *Paths to Closure,* are more technically sound and only include projected performance enhancements (productivity improvements) that can be documented. Management-related data such as disposition maps, critical closure paths, and programmatic risk assignments have been incorporated to enhance the rigor, quality, and realism of the planning process. Such data will continue to be refined.

Stakeholder, Regulator, and Tribal Nation Involvement

EM Headquarters received 39 letters during the draft *Paths to Closure* comment period, which included over 260 comments on a broad range of subjects from stakeholders, regulators, and Tribal Nations. Many of these comments were supportive of the goals and strategies outlined in the draft of *Paths to Closure.* These comments were divided into 13 distinct categories which capture those comments found to be similar in nature from multiple stakeholders: Relationship of *Paths to Closure* to Decision-making, Budget, Compliance, Uncertainties/Contingencies, End States/Stewardship, Safety and Health, Data Quality, Waste and Materials Disposition, Transportation, Enhanced Performance, Privatization, Technology Development, and Public Participation. Chapter 6 provides responses to comments in each of these categories. In addition, keeping with EM's commitment to respond to the issues of concern expressed in the letters, many of those comments have been addressed in the body of the document (see text box).

Addressing Stakeholder, Regulator, and Tribal Nation Comments

Comment Area	Addressed in Chapter
Relationship of *Paths to Closure* to Decision-making	1
Budget	2, 4, 5
Compliance	1, 4
Uncertainties/Contingencies	1, 4
End States/Stewardship	1, 3, E
Safety and Health	1, 4
Data Quality	5
Waste and Materials Disposition	1, 3, 5
Transportation	1
Enhanced Performance	4
Privatization	4
Technology Development	1, 4
Public Participation	6

The comment process was designed to give stakeholders, regulators, and Tribal Nations the opportunity to continue to participate meaningfully in the process. As these groups engage in helping to develop EM's long-term priorities and objectives, they will continue to help shape the Environmental Management program.

Chapter 1

Introduction

The Department of Energy's (DOE's) Environmental Management (EM) program has made significant progress over the past nine years in meeting the enormous challenge of cleaning up the nuclear weapons complex. Initially the program focused on characterizing waste, assessing the magnitude of contamination, stabilizing material, addressing urgent risks, and achieving compliance. Over time, EM has increased the pace at which it manages waste and cleans up sites. In 1995, EM crossed the threshold and began spending more resources on cleanup than on assessment. Now, EM can focus on completing its mission by establishing an acceleration and closure strategy. Supported by new management tools and improved estimates of the scope, schedule, and cost, EM is challenging sites to define better and more efficient ways to conduct work to achieve EM's 2006 vision (see text box).

This document, *Accelerating Cleanup: Paths to Closure* (hereinafter referred to as *Paths to Closure*), embodies stakeholder, regulator, and Tribal Nation views and comments on *Paths to Closure*. *Paths to Closure* provides:

- An integrated path forward for the management of the EM complex, based on a site-by-site, project-by-project, life-cycle foundation;
- A basis to evaluate EM's annual budgets in a long-term context;
- A response to Congressional requests for a documented management strategy for the EM program; and
- A response to concerns of stakeholders, regulators, and Tribal Nations.

> **Vision**
>
> By 2006, the Environmental Management program intends to complete cleanup at most of its 53 remaining sites. At the 10 remaining sites, including our five largest sites, treatment will continue for the remaining "legacy" waste streams. This vision will drive budget decisions, the sequencing of projects, and the actions needed to meet program objectives. This vision will be implemented in collaboration with stakeholders, regulators, and Tribal Nations.

Paths to Closure is not an action plan or a decision-making document. Furthermore, it does not show completion of EM work scope at most major EM sites by 2006. *Paths to Closure* retains a focus on 2006, which serves as a point in time around which objectives and goals are established.

Paths to Closure describes the status of EM's cleanup program and a direction forward to complete achievement of the 2006 vision. Achieving the 2006 vision results in significant benefits related to accomplishing EM program objectives. As DOE sites accelerate cleanup activities, risks to public health, the environment, and worker safety and health are all reduced. Finding more efficient ways to conduct work can result in making compliance with applicable environmental requirements easier to achieve. Finally, as cleanup activities at sites are completed, the EM program can focus attention and resources on the small number of sites with more complex cleanup challenges.

1.1 Overview of *Paths to Closure*

Paths to Closure is the Environmental Management program's[2] blueprint for completing the cleanup of contaminated soil, groundwater, and facilities; treating, storing, and disposing of waste; and effectively managing nuclear

Paths to Closure Is...	*Paths to Closure* Is Not...	Consequences
...a blueprint for EM's cleanup program.	...a decision document.	EM will make specific decisions—the need for which *Paths to Closure* identifies—following the legislative requirements of NEPA, CERCLA, RCRA, and other applicable statutes.
	...a budget document.	EM will use *Paths to Closure* to formulate annual budget strategies in the context of life-cycle cleanup costs and schedules.
...a management tool for the EM program with site-developed detailed scope, schedule, and cost data by project.	...a life-cycle cost study.	EM will use *Paths to Closure* to manage its cleanup program, including evaluating progress against performance metrics and project baselines. *Paths to Closure* will also satisfy 1994 National Defense Authorization Act reporting requirements.
...an annual account of an ongoing process.	...a one-time report.	EM plans to publish an annual *Paths to Closure* update that reflects changes made during the course of each year.

[2]Throughout this document, the phrase "Environmental Management program" or "EM program" refers to operations at both the Headquarters and site level. Section 1.3 explains the relationships of Headquarters and site levels in the EM program.

materials and spent nuclear fuel. The blueprint contains detailed site-developed scope, schedules, and costs for completing the work. Further, the blueprint identifies future decisions that must be made and defines the degree of technical and scope uncertainties.

Paths to Closure should be viewed as a management tool that reflects individual sites' best judgment as to what can be accomplished, assuming a constant funding level over time. This tool allows the EM program to formulate annual budget priorities and goals in the context of effects on life-cycle cleanup costs and schedules. The EM program recognizes that, in any given year, there will be differences between actual budget requests and the funding amount assumed in *Paths to Closure*. Such differences are inevitable because of the dynamic nature of the budget formulation process. Nevertheless, *Paths to Closure*'s role to inform annual budget deliberations is valuable because the normal range of annual budget variation is small compared with the overall life-cycle costs of the cleanup program. *Paths to Closure* will be updated annually, and these updates will allow the EM program to use the information set forth in *Paths to Closure* to assist in reviewing budget options and developing the budget. An additional benefit of the annual update is that, because it portrays the life-cycle scope, schedule, and cost for the EM program, it can meet the reporting requirements under the 1994 National Defense Authorization Act.[3]

In *Paths to Closure*, EM decided to utilize a single funding guideline and to include only those enhanced performances that sites could document in baselines. For the development of *Paths to Closure*, sites received a total funding guideline of $5.75 billion per year, which is consistent with recent appropriations. In some cases, sites exceeded this funding guideline in order to meet compliance commitments. Site funding requirements vary from year to year, as displayed in Exhibit 4-2 later in this document.

A variety of factors significantly affect the estimated scope, schedule, and cost of the EM program. Factors such as acceptance of additional facilities into the EM program, application of new technologies, or revisions of regulations, can change over time, altering the assumptions under which the EM program is conducted. To develop a foundation for estimating the scope, schedule, and cost of the program, *Paths to Closure* is based on several key planning assumptions (see text box on following page). With respect to the assumption for the Waste Isolation Pilot Plant (WIPP), the U.S. Environmental Protection Agency (EPA) has determined that WIPP can safely contain transuranic waste and that it will comply with the Agency's radioactive waste disposal standards. On May 13, 1998, the Secretary of Energy made the decision that WIPP is ready to begin disposal operations after the 30-day Congressionally mandated notification period. However, transportation of transuranic waste will be limited to non-mixed waste until the State of New Mexico has issued a Resource Conservation and Recovery Act (RCRA) Part B Permit.

---

[3]As contained in Section 3153 of Public Law 103-160, codified at 42 U.S. Code 7274k.

Paths to Closure Assumptions

Area	Assumption
Funding	Level funding at $5.75 billion per year (unless additional resources are required for compliance) from FY 1999 through program completion.
Facilities	A stable scope of facilities will be addressed in EM baselines.
Waste Management	After FY 2000, newly-generated waste will be the responsibility of the DOE programs that generate it.
Waste Disposal	The Waste Isolation Pilot Plant will open in FY 1998 to receive transuranic waste.
Site End State	End states will be determined by regulators with the involvement of local stakeholders.

Paths to Closure represents a snapshot of a single point in time in EM's cleanup program. However, the dynamic nature of the program will allow subsequent versions of *Paths to Closure* to reflect revised programmatic assumptions based upon new compliance agreements; the results of analyses prepared under the National Environmental Policy Act (NEPA); Records of Decision signed under the Comprehensive Environmental Response, Compensation, and Liability Act (CERCLA); and Statements of Basis, Closure and Post-Closure Plans, and Permits agreed to under the Resource Conservation and Recovery Act (RCRA). In addition, planned annual updates of this report will reflect cleanup progress, advances in technologies, projected savings due to demonstrated enhanced performance, the effects of annual budget allocations, and changes in site end states.

Defining end states is a key aspect of defining the scope of the cleanup program. Once the end state of a site is known, the work necessary to achieve that end state can be divided into steps, and the steps can be organized in an appropriate sequence. Currently, *Paths to Closure* is based on the best available end state assumptions (i.e., planned end points) made by each site with respect to EM activities. However, decisions about end states and cleanup approaches to achieve those end states will be made in accordance with the requirements of CERCLA, RCRA, and other applicable statutes (with appropriate environmental review) and may differ from the assumptions described in this document. It should also be noted that the

> **A Site is Considered "Complete" (or at its End State) When...**
>
> - Deactivation or decommissioning of all facilities currently in the EM program has been completed, excluding any long-term surveillance and monitoring;
> - All releases to the environment have been cleaned up in accordance with agreed-upon cleanup standards;
> - Groundwater contamination has been contained, or long-term treatment or monitoring is in place;
> - Nuclear material and spent fuel have been stabilized and/or placed in safe long-term storage; and
> - "Legacy" waste (i.e., waste produced by past nuclear weapons production activities, with the exception of high-level waste) has been disposed of in an approved manner.

completion of cleanup activities at many sites (see text box) does not mean there will no longer be an EM presence at the site. Many sites will require additional surveillance and monitoring funded by EM, and some will have an ongoing, non-EM mission, such as research and development not related to environmental matters.

Current site assumptions about planned end state do not rule out future decisions to clean up a site to a different end state from that envisioned under those assumptions. In fact, site versions of *Paths to Closure* explicitly state that the planning end point assumed for purposes of establishing baselines may not represent the ultimate end state of any given site. Improvements in end states may be possible at some time in the future with the development of new technologies, more economical cleanup approaches, the availability of additional resources, and/or changes in the interests of stakeholders, regulators, and Tribal Nations.

The EM program is developing an integrated management system to align more closely three aspects of its efforts: life-cycle planning, the annual budget formulation process, and the measurement of results. To facilitate that objective, the EM program organized all cleanup activities into discrete projects. For the first time, an integrated life-cycle database has been developed to maintain information about those projects. The process of establishing specific projects and baselines with scope, schedule, and costs has resulted in significant reductions in EM life-cycle cost estimates since the initiation of the cleanup strategy in 1996.

1.2 Background on the EM Program and Mission

During the past nine years, the EM program has grown from infancy to its present status as a major focus of DOE. This section provides a brief description of the EM program, its history, and the current context of its efforts to pursue the *Paths to Closure* vision.

1.2.1 What is the Environmental Management Program?

During the Cold War period of nuclear weapons production, awareness of the effects of environmental pollution grew significantly. Congress enacted a series of stringent environmental protection laws that empower both federal and state regulatory agencies to oversee federal activities affecting the environment. In 1989, DOE established the EM program to address the contamination and waste created by nuclear weapons production, research, and testing activities during the Manhattan Project and the Cold War era in a manner consistent with applicable environmental laws. Those activities included mining and milling of uranium, uranium enrichment, fuel and target fabrication, reactor operations, chemical separations, weapons component fabrication, weapons operations, and research, development, and testing.

The primary mission of the EM program is to reduce threats to health and safety posed by contamination and waste (referred to as "legacy" activities or problems) at DOE sites including those associated with the nuclear weapons complex. EM's mission is realized through the following program areas: waste management; stabilization of nuclear material and spent fuel; deactivation and decommissioning of facilities; remedial actions to soil and water; infrastructure and support; and national programs focused on such activities as science and technology development, transportation, emergency management, and pollution prevention.

The EM program manages its cleanup work through 11 Operations/Field Offices across the United States. Offices are located in the following areas: Albuquerque, New Mexico; Carlsbad, New Mexico[4]; Chicago, Illinois; Idaho Falls, Idaho; Las Vegas, Nevada; Oakland, California; Oak Ridge, Tennessee; Miamisburg, Ohio; Richland, Washington; Jefferson County, Colorado; and Aiken, South Carolina. Each Operations/Field Office is responsible for cleanup activities at one or several sites. The EM program historically has identified 134 "geographic sites" (distinct geographic locations that generated waste or were contaminated by DOE or predecessor agency activities) as part of its scope. These sites are located in 31 states and one territory and encompass an area of over two million acres—equal to the size of Rhode Island and Delaware combined. At the beginning of 1998, cleanup responsibility for 21 sites managed by EM under the Formerly Utilized Sites' Remedial Action Program (FUSRAP) was transferred to the U.S. Army Corps of Engineers. *Paths to Closure* addresses the remaining 113 sites, including required long-term surveillance and monitoring of the 60 sites completed before FY 1998 and environmental management activities for 53 additional sites. Appendix C contains a complete list of sites and completion dates.

[4]Technically, Carlsbad is an Area Office; however it is included in discussions of Operations/Field Offices throughout this report.

1.2.2 *Historical Management: From the Cold War to Environmental Cleanup*

The threat to national security initiated during World War II led to the development of a substantial, high-security engineering and production operation. Over the past five decades, DOE and its predecessor agencies developed the largest government-owned industry in the United States. This entity was responsible for the research, development, testing, and production of nuclear weapons and a variety of nuclear-related research projects. To protect national security interests, information on these activities was generally limited to a small group of managers, researchers, and workers and was generally kept from public knowledge.

During the Cold War era, the relatively unconstrained availability of resources fostered "level-of-effort" management approaches such as contracting for the full-time commitment of an agreed-upon number of personnel rather than for the accomplishment of specific tasks in specified time frames. Moving the focus of DOE's effort from production to cleanup required that the management and organizational culture move away from the "level-of-effort" approach towards a more open, project-oriented cleanup program in which stakeholders would have effective involvement. After a 50-year operating history, the effort required to make these changes was significant. The abrupt end of the Cold War in the late 1980's also brought an end to the availability of relatively unbounded resources.

Now, the EM program must focus on completing cleanup through the adoption of management strategies based on project needs. The EM program must increase its public accountability, committing itself to public involvement throughout the cleanup process. Further, the EM program must complete its cleanup activities with stabilized funding and staffing levels, while demonstrating measurable progress. All the while, EM must maintain its focus on safety and health and regulatory compliance.

> **Understanding the Legacy**
>
> Through publications such as *Closing the Circle on the Splitting of the Atom,* the *Baseline Environmental Management Report, Taking Stock, Linking Legacies,* and now *Paths to Closure,* the EM program has worked to inform the public about the past, present, and future of the nuclear weapons complex and resulting cleanup activities. (See Appendix F, List of References)

1.3 Relationship of *Paths to Closure* to the EM Decision-making Process

Public comments on the February 1998 draft *Paths to Closure* requested clarification on the decision-making process for the work described in *Paths to Closure*. Decisions in the EM program are driven by various statutory mandates, most notably CERCLA and RCRA. Most decisions are made at the site level (with appropriate Headquarters oversight). Other decisions are made at the Headquarters level because of their complex-wide implications. In many cases, ultimate decision-making authority, in the sense of final approval authority, resides with EPA or state regulators.

Public participation is an important element of the EM program's decision-making process. NEPA requires federal agencies to consider the environmental impacts of their proposed actions. NEPA also requires that the public be informed of, and have an opportunity to comment on, major federal actions significantly affecting the environment. Consistent with its obligations under NEPA, the EM program performs an appropriate level of environmental review in connection with its projects, with opportunities for public involvement. For projects managed under CERCLA, EM relies on the CERCLA process to incorporate NEPA values.

Paths to Closure outlines EM's current estimate of the scope, schedule, and costs for each site to complete the cleanup program. The estimate includes projects for which key site cleanup decisions have been made pursuant to CERCLA, RCRA, or other statutes, and projects where such decisions have yet to be made. Where decisions have not yet been made, sites make **assumptions** (e.g., site planning end states) about how those cleanup actions might be carried out so that sites can define work and develop schedule and cost estimates. In those cases where decisions have not yet been made, the Environmental Management program will follow the decision-making processes called for by the relevant statutory authority that governs the activity in question (e.g., CERCLA or RCRA) with appropriate environmental review.

Paths to Closure also includes cost estimates for federal salaries, investments in science and technology development, and miscellaneous support functions. EM sites and EM Headquarters make decisions through the budgetary process on the scope and pace of work for these activities. Stakeholders and Tribal Nations will have significant opportunities to be involved in all decision-making processes.

1.3.1 EM Decision-making Processes

EM projects typically consist of six phases:

(1) Planning, where initial project planning occurs;

(2) Study, where projects are characterized and alternative solutions are evaluated;

(3) **Recommendation**, where a preferred solution is identified;

(4) **Decision**, where a formal decision is made;

(5) **Implementation**, where the work to execute the decision is conducted; and

(6) **Monitoring**, where actions taken during project implementation are maintained.

The names of these project phases may differ by statute. For example, in CERCLA, the study phase is called a Remedial Investigation, while under RCRA it is called a RCRA Facility Investigation. Conceptually, however, the study phases of projects conducted under each of the different statutes are analogous to one another. Similarly, other phases of projects conducted under different statutes are analogous to each other, even if the terminology is different.

EPA or state environmental regulators are the final decision-makers for cleanup work conducted under CERCLA and RCRA because of their regulatory approval roles. At the site level, the Environmental Management program negotiates with state and federal regulators regarding the scope and schedule for conducting the studies, confers with the regulators on the recommended course of actions, and negotiates with the regulators on the scope and schedule for implementing and monitoring the actions once decisions have been finalized. The EM program's role is to comply with schedules negotiated with state and federal regulators for conducting studies, proposing recommended courses of action, and implementing the actions once the regulators have made decisions.

For work performed as a result of decisions informed by the NEPA process, EM makes decisions in accordance with the Council on Environmental Quality's regulations implementing NEPA and the Department's own NEPA-implementation regulations.

1.3.2 Paths to Closure *Relationships*

Exhibit 1-1 illustrates a conceptual decision-making process applicable to CERCLA, RCRA, NEPA, or any other statutory framework, and the relationship of *Paths to Closure* to that process. As the exhibit illustrates, projects advance through the decision process over time. As a project (or project activity) moves through the stages, additional information is collected. Therefore, the uncertainty about project scope, costs, and schedule of the implementation phase diminishes as indicated by the length of the dotted arrows in Exhibit 1-1.

Because each yearly version of *Paths to Closure* is a vantage from a single point in time, EM makes a series of evolving planning assumptions about future activities based on information generated and decisions made during the previous year. As mentioned above, assumptions about specific projects do not bias decisions that will be made about those projects, nor do they eliminate or restrict alternative approaches or opportunities for public involvement in the decision-making process.

Exhibit 1-1

Relationship of EM Project Decision-making Process to *Paths to Closure* Data

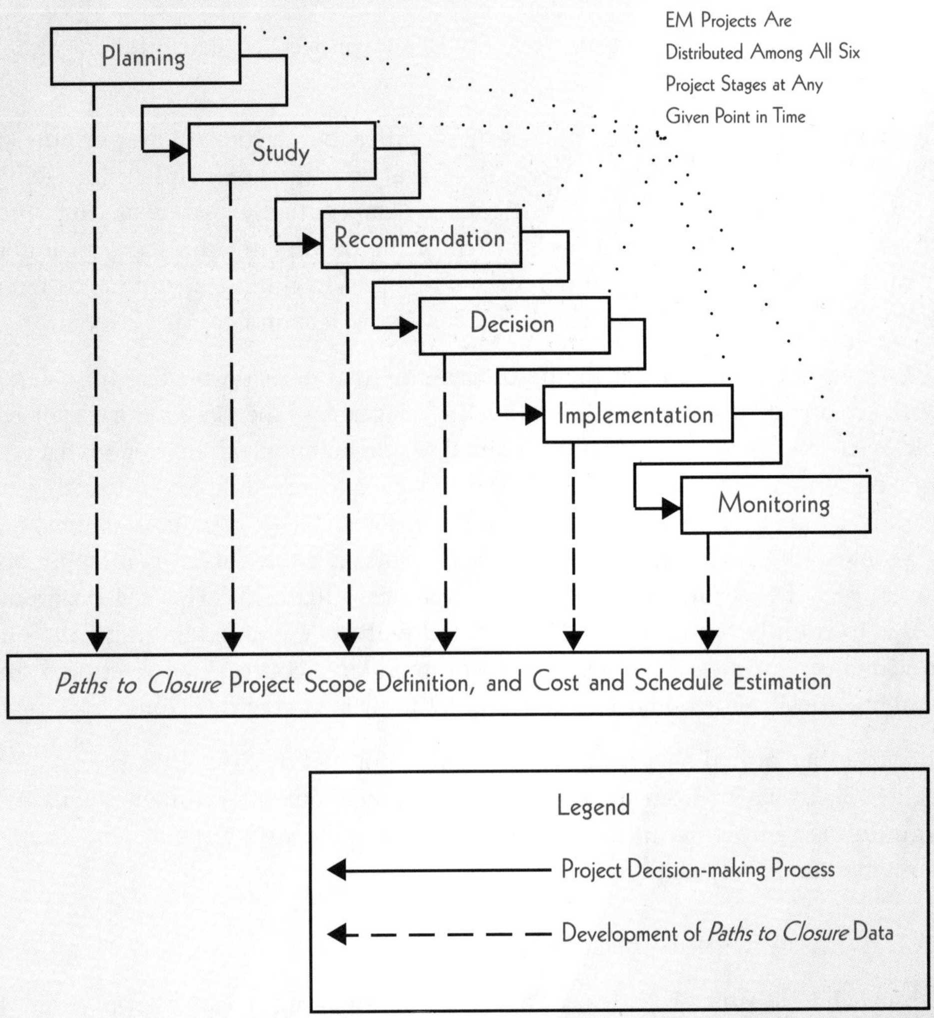

Paths to Closure also identifies opportunities to accelerate the pace of projects or parts of projects made under CERCLA, RCRA, and NEPA, such as the completion of specific cleanup projects more rapidly than may be required under compliance agreements or the pace at which EM performs environmental impact statements. We will ensure that acceleration of the pace of cleanup activities does not reduce cleanup scope and does not compromise the health and safety of workers or the achievement of appropriate cleanup standards.

In addition, *Paths to Closure* plays an important role in EM's site and Headquarters budget processes. Sites use their *Paths to Closure* reports as a guide to developing site budget priorities. EM Headquarters uses *Paths to Closure* to formulate annual budget strategies in the context of life-cycle cleanup costs and schedules. *Paths to Closure* is also a useful tool for making annual adjustments to the execution of the cleanup program based on budget funding decisions. Chapter 4 describes the relationship of *Paths to Closure* to the budget process in greater detail.

1.4 Safety and Health and Regulatory Compliance

Since its inception, the EM program has placed a high priority on achieving its mission in a manner that ensures a safe and healthy workplace, reduces risk, and attains compliance with all applicable regulatory requirements. *Paths to Closure* embraces those objectives in accelerating cleanup efforts. However, comments of stakeholders, regulators, and Tribal Nations on the *Discussion Draft* expressed concern that initial development of *Paths to Closure* had focused on defining the scope, schedule, and cost of the cleanup at the perceived expense of these cleanup objectives.

1.4.1 Safety and Health

A fundamental objective of the EM program is to ensure the protection of workers and the public throughout the conduct of its cleanup mission. The EM program's cleanup workers, including federal employees, contractors, and subcontractors, are the most vulnerable to hazardous exposure and risk. Such workers are frequently engaged in activities that involve radioactive and toxic wastes, and under conditions that are conducive to industrial accidents. The EM program has a responsibility to protect the safety of its workers; failure to meet that responsibility is unacceptable.

That philosophy is reflected in EM's safety and health policy: "Do Work Safely or Don't Do It." The need to accelerate cleanup and reduce costs does not alter that commitment to safety. In implementing the project-oriented approach presented in *Paths to Closure*, protection of worker health and safety is built into each specific project across the complex. The Environmental Management program is implementing the principles of Integrated Safety Management in all projects so that safety and health become an integral part of project management. That approach is consistent with the best in industry, and it reduces accidents and improves work planning. Those benefits may in turn give rise to performance enhancements through reductions in workers compensation premiums, reduced lost productive time, and enhancements in work planning and execution.

EM's safety and health activities, therefore, become an integral component of EM's planning, budgeting, and accountability management system. In addition, reducing risk to workers, the public, and the environment is an integral element of EM's approach to setting priorities, sequencing project work, and measuring performance. Efforts to accelerate activities can in turn accelerate risk reduction. Initiatives set forth in *Paths to Closure* place priority on projects that eliminate urgent risks.

1.4.2 *Regulatory Compliance*

The EM program will comply with all activities required under applicable federal, state, and local environmental statutes and regulations; activities required under the terms of permits, administrative orders, or judicial decrees; enforceable milestones or schedules established in agreements negotiated between EM and its regulators; and commitments to the Defense Nuclear Facilities Safety Board (DNFSB). All site versions of *Paths to Closure* reflect and explicitly state this position. To support this position, Operations/Field Offices are required to identify regulatory drivers for projects as well as all significant enforceable agreement milestones. Additionally, all Operations/Field Office budget requests must include an integrated project priority list which is tied to regulatory compliance drivers. EM's commitment to compliance is discussed further in Chapter 4.

1.5 Easing the Transition of Workers

Workforce restructuring plans are currently in place or under development for the sites that will address adjustments in the workforce that may occur from time to time as cleanup activities are completed at a site. Potential strategies for offering benefits to workers affected by workforce adjustments are under review. These strategies are focusing on approaches that are linked to requirements identified by a comprehensive personnel resource management plan. They may include incentive programs for both voluntary and involuntary separation and outplacement assistant services, such as job search workshops, access to job listings, resume preparation, career and educational counseling, and educational assistance to help workers make the transition to new job opportunities. Certain involuntarily separated workers will be eligible for preference in hiring and for severance pay, in accordance with Section 3161 of the National Defense Authorization Act for FY 1993. Some approaches may include providing benefits prior to employee separation.

As projects come to a close and sites approach closure, DOE also intends to provide, in accordance with Section 3161 of the National Defense Authorization Act for FY 1993, assistance to communities that are affected by the reconfiguring, downsizing, and closing of its defense nuclear facilities. DOE realizes that attaining *Paths to Closure* goals may affect the economies of nearby communities where a significant number of displaced workers live. DOE will cooperate with the Community Reuse Organization and execute economic development initiatives to help minimize those effects. The Office of Worker and Community Transition, which is responsible for the overall management of DOE's community transition program, will authorize specific actions, within approved funding levels, selected through application of the evaluation criteria set forth in the guidance.

The remainder of this report is organized into five chapters and a series of appendices. ***Chapter 2*** *summarizes the scope, schedule, and costs for the Environmental Management cleanup program.* ***Chapter 3*** *provides more detailed scope, schedule, and cost information for three Operations/Field Offices: Rocky Flats, Richland, and Savannah River. (****Appendix E*** *provides analogous information for the remaining eight Operations/Field Offices.)* ***Chapter 4*** *discusses EM efforts to meet programmatic challenges, largely focusing on mechanisms to accelerate cleanup and reduce costs.* ***Chapter 5*** *describes the new integrated system EM intends to use to manage the cleanup program.* ***Chapter 6*** *summarizes stakeholder, regulator, and Tribal Nation comments and EM program responses to comments on the February 1998 draft Paths to Closure.*

Chapter 2

Baseline Scope, Schedule, and Cost

Chapter 2 presents the scope, schedule, and cost of the Environmental Management (EM) cleanup program. This chapter begins with a discussion of the approach taken by sites to the development of baselines and the relationship of those baselines to the Project Baseline Summaries (PBSs) used to aggregate the data in *Paths to Closure*. Following the discussion on baselines, the chapter provides a summary of the baselines for each Operations/Field Office, a profile for the completion of Environmental Management work at each site, a discussion of how the EM program is managing its cleanup schedule and a reconciliation with the Department's FY 1997 Financial Statement. The basic work scope, cost, and schedule data in this report has not changed since the publication of the February draft *Paths to Closure*.

2.1 The Development of Site Baselines

One of the fundamental improvements to the management of the EM program is the aggregation of units of work essential to EM's cleanup mission into projects. The creation of projects enables Field managers to develop detailed projections of scope, schedule, and cost (that is, a baseline) for each site, based upon the aggregation of logical, discrete units of work. Historically, during the nuclear weapons production phase, sites used mostly level-of-effort methodologies to develop estimates. In contrast, site baselines, built from individual project baselines, are the foundation for cost projections in *Paths to Closure*. The direct link of scope, schedule, and cost estimates in site baselines to estimates in *Paths to Closure* means that the quality of data in the document is linked directly to the quality of site baselines.

One key determinant of quality is the definition of scope. It is more difficult to develop a baseline for a technically challenging, first-of-its-kind project than for a clearly-defined project that is based on an established approach. The EM program is responsible for a massive environmental cleanup effort, much of which is the first of its kind. A good example of the type of challenge that the Environmental Management program faces is the cleanup of high-level waste tanks at the Hanford Site, a project which is estimated to cost $30 billion (constant 1998 dollars) over the life cycle. The Hanford high-level waste project has been characterized as one of the most challenging engineering projects ever undertaken. Given the technological challenges and the uncertainties involved with the characterization of tank waste, the chemical interactivity of the constituents, the method of removal of waste from the bottom of the tanks, and the processing method that will be applied once the material has been removed from the tanks, the overall baseline for this project encompasses a great deal of uncertainty.

Despite uncertainties, EM's knowledge has increased substantially over the past several years, supporting the development of better baselines. The development of conceptual approaches to the storage, treatment, and disposal of all waste types at all sites is an example of the progress that the EM program has made. Such conceptual approaches, reflected in schematic diagrams called **disposition maps**, provide a picture of the scope of the EM program's environmental restoration and waste management activities. In addition, the maps simultaneously identify uncertainty related to overall scope and disposition. Each site also has improved its understanding of its **critical closure path**, that is, the universe of activities that must be completed on time in order for EM activities to be completed as scheduled. Disposition maps and critical closure paths are works in progress that help document the scope, schedule, and cost of the EM program at each site. A short-term priority for the EM program is to continue to improve its understanding of the scope of the cleanup program through the refinement of baselines and related tools, including disposition maps and critical closure paths.

As part of the overall guidance for developing baselines, sites were given a funding guideline of $5.75 billion per year, which is consistent with recent appropriations. Some site baselines currently exceed their share of the $5.75 billion per year funding guideline to show compliance requirements. In response to concerns expressed by stakeholders, regulators, and Tribal Nations, the EM program requested that the sites include assumptions of enhanced performance (reductions in cost achieved through increased efficiency), integration assumptions, and other cost-saving assumptions only in cases in which sites were confident that such performance could be demonstrated or where stakeholders, regulators, and Tribal Nations have approved them.

Sites provided information from their baselines to support *Paths to Closure*, primarily in the form of PBSs. Appendix A presents a complete list of PBSs. A **PBS** is not the project baseline, but rather a management tool that summarizes information about each project (see text box). PBSs are used for planning, budgeting, and evaluation. Appendix B provides a sample PBS.

Key Elements of a Project Baseline Summary

- Scope
- Schedule
- Cost
- Risk
- Technical Approach
- End State
- Regulatory Drivers
- Safety and Health
- Performance Metrics

2.2 Operations/Field Office Estimates of Cost and Closure

The PBS for each project includes information about scope, schedule, and cost from 1997 through 2070. While all EM cleanup activities are scheduled for completion before 2070, some long-term surveillance and monitoring and stewardship activities will continue beyond 2070. *Paths to Closure*, however, includes only costs through 2070. In each PBS, Operations/Field Offices reported costs in current year dollars; therefore, the cost estimates have already been adjusted for inflation (assumed to be 2.7 percent per year) and indicate the cost at the expected time of the outlay. Inflation lowers the "buying power" of each dollar over time, so a project that costs $5 million current year dollars in 1998 is more expensive, in relative terms, than a project that costs $5 million in current year dollars in 2006. The use of constant 1998 dollars in discussions of cost estimates in *Paths to Closure* ensures the comparability of costs over time, eliminating those variations that are the result solely of inflation.

The EM program baseline is based on 353 PBSs. The cost estimate (1997 through 2070) for the EM program—$147.3 billion in constant FY 1998 dollars—aggregates costs for all 353 PBSs. Exhibit 2-1 shows the overall estimate by Operations/Field Office. The 53 sites in the "Number of Sites Completed" columns include sites planned for completion in 1998 and beyond. Historically, 60 sites were completed through 1997. Appendix C provides a complete list of geographic sites with their actual or planned completion dates.

Exhibit 2-1 shows that the current site baselines support the 2006 vision of completing cleanup at most sites by 2006. However, it also shows that by 2006, completion of EM activities occurs primarily at the Department's smaller sites. After 2006, EM's greatest challenge will be to complete cleanup at some of the largest and most technically complex sites. In fact, 77 percent of the estimated costs after 2006 are accounted for by the Savannah River Site, the Hanford Site (managed by Richland), and the Idaho National Engineering and Environmental Laboratory.

Exhibit 2-1
EM Costs by Operations/Field Office

Operations/ Field Office	Estimated EM Costs (1997-2006)	Estimated EM Costs (2007-2070)	Total Estimated EM Costs (1997-2070)	Number of Sites Completed	
	(All costs in billions of constant 1998 dollars)			1998-2006	After 2006
Albuquerque	2.1	2.0	4.1	12	1
Carlsbad[a]	1.8	5.9	7.7	0	1
Chicago	0.3	0.0	0.3	5	0
Headquarters/ National Programs	5.7	5.6	11.3	NA	NA
Idaho	5.0	11.3	16.3	0	1
Nevada	0.9	1.3	2.2	8	2
Oakland	0.7	0.3	1.0	8	1
Oak Ridge	5.4	7.7	13.1	3	2
Ohio	4.6	0.2	4.8	5	1[b]
Richland	13.0	37.3	50.3	0	1
Rocky Flats	5.3	1.0	6.3	0	1[c]
Savannah River	12.0	17.7	29.7	0	1
TOTAL[d]	57.0	90.3	147.3	41[e]	12
				53	

[a] Costs for the Carlsbad Area Office include the costs associated with operating the Waste Isolation Pilot Plant as the national repository for the disposal of transuranic waste and the costs of decommissioning the site after disposal operations have ended.

[b] The one site after 2006 is the Fernald Environmental Management Project (FEMP). It is expected that cleanup at FEMP also will be completed before 2006, although the baseline currently indicates completion in 2008.

[c] The current baseline for the Rocky Flats Environmental Technology Site reflects a 2010 closure. However, the baseline is being revised to reflect the commitment to complete closure by 2006.

[d] Individual costs may not sum to totals due to rounding.

[e] With the accelerated goal of cleaning up the Rocky Flats Environmental Technology Site and the Fernald Environmental Management Project (by 2006 and 2005 respectively), the number of sites completed by 2006 would be 43.

Exhibit 2-2 displays the life-cycle cleanup costs of the EM program, over time, by Operations/Field Office. "Other Operations/Field Offices" in Exhibit 2-2 includes Albuquerque, Carlsbad, Chicago, Headquarters/National Programs, Nevada, and Oakland.

Exhibit 2-2
Environmental Management Cleanup Costs by Operations/Field Office Over Time

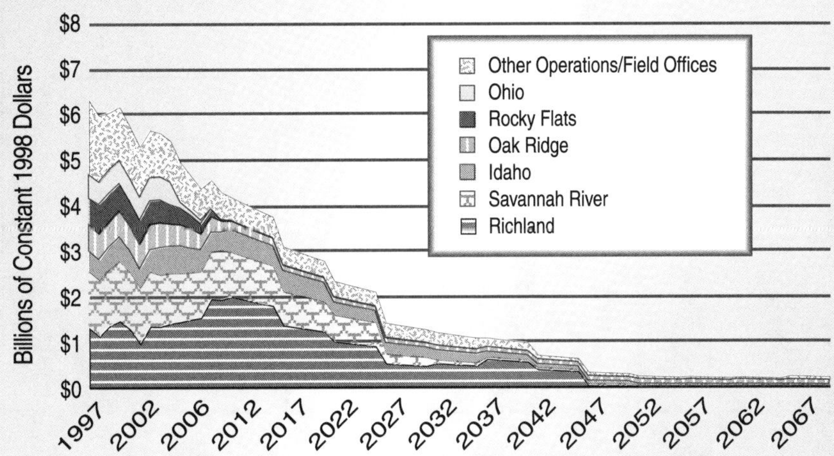

2.3 Details of Life-cycle Costs

This section presents details of the life-cycle cleanup costs for the EM program. First, the section relates costs to the types of work EM performs, thereby outlining major cost drivers for the program. Second, the section breaks out EM costs by state. Third, the section explains other scope and costs that, while not the focus of *Paths to Closure*, are nevertheless important to put this report in context. Finally, the section displays costs by a system of categories that parallels EM's current budget structure, shows the benefits of aggregating units of work into projects, and illustrates the EM program's focus on the completion of specific projects by 2006.

2.3.1 Cost by Category of the EM Work Scope

The $147.3 billion life-cycle cost estimate includes the costs of completing all known EM work scope. To provide additional insights on cost, each Operations/ Field Office estimated the distribution of costs by scope category. These supplementary data by category are presented in Exhibit 2-3. Brief explanations of the categories follow the exhibit.

Exhibit 2-3
Life-Cycle Costs by Category

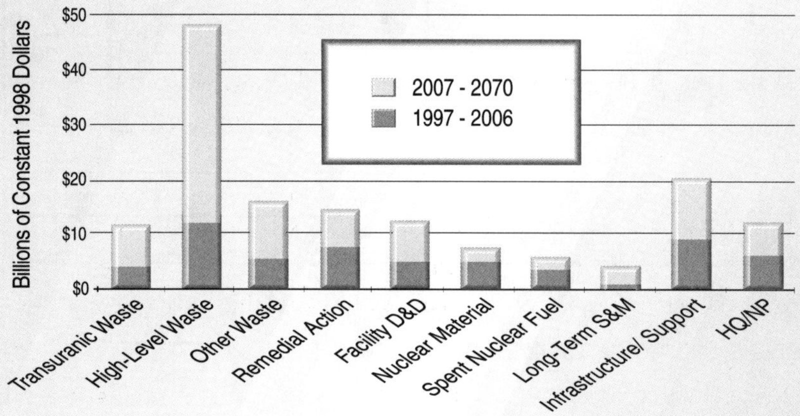

High-level Waste. Currently, the EM program is responsible for the storage, treatment, and stabilization of hundreds of thousands of cubic meters of highly radioactive waste generated from decades of nuclear weapons production, mostly at the Savannah River Site, the Hanford Site, and the Idaho National Engineering and Environmental Laboratory. High-level waste also is found at the West Valley site in New York. High-level waste management is by far the largest cost driver for EM; it is estimated to account for 32 percent of the total cost of the EM program over the life cycle. Approximately 74 percent of these costs will remain after 2006.

Transuranic Waste. The EM program is responsible for the storage, treatment, and disposal of approximately 130,000 cubic meters of contact- and remote-handled transuranic waste from known defense-related testing and experimental projects. This estimate includes the volume of transuranic waste that is currently stored and that which is expected to be generated. The EM program expects to dispose of an additional 40,000 cubic meters of such waste generated from continuing and future missions as well as decommissioning and other defense-related projects of DOE. Before it can be shipped, transuranic waste requires safe storage and sometimes requires treatment. Currently, transuranic waste activities are estimated to be seven percent of the total cost of the EM program through 2070. Sixty-six percent of the cost for transuranic waste will be incurred after 2006.

Other Waste. The EM program must manage millions of cubic meters of other types of waste including low-level radioactive waste, hazardous waste, and mixed low-level waste (containing both radioactive and hazardous constituents). Some of that waste is in storage awaiting treatment and disposal; more such waste will be generated during the cleanup process. Virtually all sites manage one or more of these types of waste. The EM program currently is estimating that 11 percent of its total cost will go toward addressing these types of waste over the life cycle.

Remedial Action. The EM program is responsible for characterization and cleanup of approximately 9,000 "release sites." A release site is a specific area, within a larger geographic site, at which contaminants or contaminated materials might have been spilled, dumped, disposed of, or abandoned. The cleanup of release sites involves the remediation of soil, surface water, and/or groundwater. Some release sites require no further action while others require remediation or monitoring. Release sites range in size from very small spills to large dumping areas. Currently, it is estimated that 80 percent of the release sites will be cleaned up by 2006. Characterization and remediation of release sites are estimated to account for 10 percent of the total cost of the EM program over the life cycle.

Facilities. EM's facilities range from small guardhouses to massive excess production facilities and nuclear reactors. Combined, the area of these facilities currently assigned to EM is more than 65 million square feet. This total square footage exceeds the area of 1,300 football fields. Most of the large buildings contain contaminated equipment, machinery, and pipes. Others store waste and nuclear materials. Most of the buildings require deactivation, decontamination, and decommissioning. These facilities are projected to account for eight percent of the total cost of the EM program over the life cycle.

Nuclear Materials. Nuclear materials include plutonium, uranium, and other materials in various forms (for example, metals, oxides, solutions, residues). These materials need to be stabilized and prepared for their ultimate disposition. EM plans to complete most of this work by 2006. The EM program anticipates that four percent of the total life-cycle cost of the EM program will be incurred by the stabilization, packaging, and management of nuclear materials.

Spent Nuclear Fuel. Spent nuclear fuel includes fuel, targets (excluding medical isotope targets), slugs, and sludge. The Idaho National Engineering and Environmental Laboratory, the Savannah River Site, and the Hanford Site generated most of the existing spent nuclear fuel. The EM program also manages foreign research reactor spent fuel. The EM program estimates that three percent of the total Environmental Management cost over the life cycle will go toward spent nuclear fuel management. Most stabilization activities are scheduled for completion by 2006.

Long-term Surveillance and Monitoring. The Environmental Management program is responsible for the long-term surveillance and monitoring of up to 81 sites. Surveillance and monitoring activities currently account for three percent of the life-cycle estimate. However, some sites need to further refine estimates in this area. A site is considered to be complete before long-term surveillance and monitoring activities end; at some sites these activities will continue well beyond 2070.

Infrastructure and Support. The Environmental Management program maintains site infrastructure, conducts program management and oversight activities, and manages other efforts to ensure the safety and health of workers and the public and to protect the environment while conducting cleanup activities. At some sites, the EM program provides such services as utilities, security, road maintenance, facilities upgrades, and similar activities. The EM program estimates that 14 percent of its total life-cycle costs will be allocated to these activities. At some sites, these costs are allocated to specific waste management or remedial action activities. Therefore, some infrastructure/ support costs are captured in other categories.

National Programs and Headquarters. This category includes program direction, which funds federal salaries and related costs for the entire EM complex (both Headquarters and the Field). National programs include such crosscutting projects as the National Transportation program, the National Pollution Prevention program, and the National Science and Technology program. The EM program expects that eight percent of its life-cycle costs will be expended on these activities.

2.3.2 Cost by State

As of the beginning of FY 1998, there were 53 sites in the EM program that still require cleanup and associated funding. EM will also continue to require funding for activities at other sites (such as long-term surveillance and monitoring for completed sites) and some amount for federal salaries at both Headquarters and in the Field. Exhibit 2-4 outlines the estimated costs of the EM program by state.

2.3.3 Other Scope and Costs

End state assumptions (i.e., assumed end points) in *Paths to Closure* differ from those made in previous EM life-cycle cost estimates to reflect current site end state assumptions. For example, *Paths to Closure* does not include the costs associated with decommissioning the Portsmouth Gaseous Diffusion Plant in Ohio and the Paducah Gaseous Diffusion Plant in Kentucky and may not include the full costs for decommissioning some facilities, such as the spent fuel pools and canyons at the Savannah River Site in South Carolina. As assumptions change,

Exhibit 2-4
Estimated EM Life-cycle Costs by State[a]

State	Estimated Cost (in billions of constant 1998 dollars)[b] 1997-2070
California	$0.8
Colorado	$6.5
Florida	$0.3
Idaho	$16.4
Illinois	$0.1
Kentucky	$0.9
Missouri	$0.4
Nevada	$2.2
New Mexico	$9.5
New York	$1.5
Ohio	$4.6
South Carolina	$29.7
Tennessee	$11.0
Texas	$0.1
Utah	$0.1
Washington	$50.4
Multiple States (Long Term S&M)	$2.3
Multiple States (Program Direction)	$7.6
Multiple States (Science and Technology Development)	$2.9
Multiple States (All Other, Including National Programs and HQ)	$0.1

[a]Other states include Alaska, Iowa, Massachusetts, Mississippi, and New Jersey.
[b]Individual costs may not sum to $147.3 billion due to rounding.

future updates to *Paths to Closure* will be adjusted accordingly. The effect of the adjustment to meet such needs could be significant. The 1996 *Baseline Environmental Management Report* estimated the cost of decommissioning such facilities at more than $10 billion.

In addition to the baseline costs outlined in Sections 2.2 and 2.3, PBSs include other costs that require explanation. *Paths to Closure* was developed under the assumption that the EM program will not accept any newly-generated, non-EM waste after FY 2000. For the Operations/Field Offices that manage those wastes, especially those that manage waste at operating national laboratories (for example, Albuquerque, Chicago, Oakland, and Oak Ridge), responsibility is expected to be transferred to the generator after FY 2000, which is usually another program of the Department, such as the Defense Programs or Energy

Research. Exhibit 2-5 shows these costs in the column labeled "Costs Transferred to Other Programs." The EM program expects to transfer EM budget target dollars associated with newly-generated, non-EM waste to the generators as well. Should this assumption change, the affected project baselines (and PBSs) will require revision.

Exhibit 2-5

EM Baseline Costs and other Costs by Operations/Field Office

Operations/ Field Office	EM Baseline Cost[a]	Costs Transferred to Other Programs	Baseline Costs Paid by Other Entities
	(billions of constant 1998 dollars)		
Albuquerque	4.1	4.5	<0.1
Carlsbad	7.7	0	0
Chicago	0.3	1.1	0
Headquarters/ National Programs	11.3	0	<0.1
Idaho	16.3	0	0
Nevada	2.2	0	0
Oakland	1.0	1.1	0
Oak Ridge	13.1	1.4	0.1
Ohio	4.8	0	0
Richland	50.3	0	0.5
Rocky Flats	6.3	0	<0.1
Savannah River	29.7	0	0.1

[a]Individual costs may not sum to $147.3 billion due to rounding.

In other cases, costs may be paid by other DOE programs or entities outside of DOE to support the cleanup at EM sites. Some examples include state contributions to the Uranium Mill Tailings Remedial Action Project and the co-funding of some EM activities with DOE's Office of Defense Programs. The EM program anticipates such funding will continue. The discussion in Section 2.2 excluded funds contributed by these other entities to cover such costs; however, such costs are shown in Exhibit 2-5 in the column labeled "Baseline Costs Paid by Other Entities." Exhibit 2-5 also displays the EM baseline cost (from Section 2.2).

Finally, the current baseline assumes that the EM program will not accept additional surplus facilities for deactivation and decommissioning. However, the Department is considering transferring additional surplus facilities to the EM program beginning in 2002 with limited exceptions occurring before that date. If and when such transfers occur, the EM program will develop projects and adjust current assumptions to account for the cleanup of these facilities and include these costs in future updates to *Paths to Closure*.

2.3.4 *Cost by Category of Project Completion Date*

For the FY 1999 budget request, the EM program developed a new categorization structure based upon the projects included in *Paths to Closure*. The new structure includes three program budget accounts:

- **Closure** includes all projects at sites closed by 2006 without a continuing DOE mission.
- **Project Completion** includes sites completed by 2006 with an ongoing DOE mission, and projects completed by 2006 at sites with cleanup work continuing after 2006.
- **Post-2006 Completion** includes projects that are expected to require work beyond FY 2006.

The new structure also identifies three additional accounts: Technology Development, Program Direction (i.e., federal salaries), and Privatization projects. Exhibit 2-6 shows the baseline cost of the EM program broken out over time into the Closure, Project Completion, and Post-2006 Completion accounts. Most of the projects in the Closure and the Project Completion accounts are scheduled for completion by 2006. Other projects and/or sites could move into project completion or closure as they achieve additional enhanced performance.

Exhibit 2-6
Baseline Cost by Closure, Project Completion, and Post-2006 Completion[a]

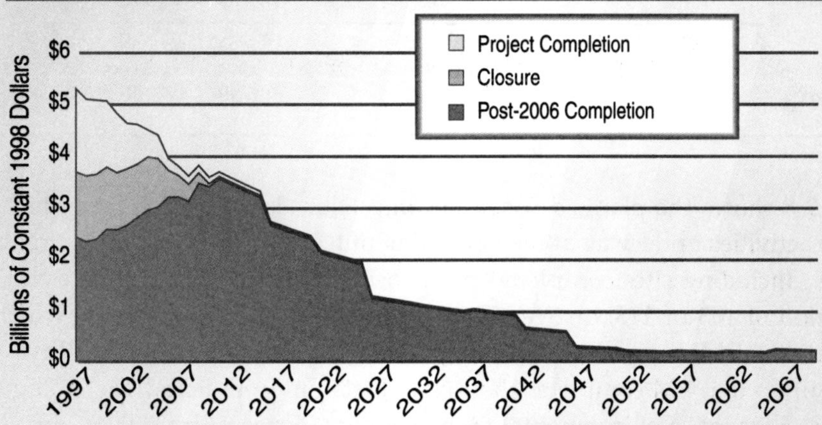

[a] Does not include technology development, program direction, or privatization. Funding beyond 2006 for projects in the Closure and Project Completion accounts is for long-term surveillance and monitoring and for the baseline (non-accelerated) closure strategy for Rocky Flats (closed 2010) and Fernald (closed 2008).

2.4 Completion Schedule for the EM Program

Each Operations/Field Office estimated a completion date for major EM activities at each site and for each of its projects. The definition of "complete," as outlined in Chapter 1, does not assume that the EM program or DOE will leave a site when cleanup activities at that site are considered complete. Instead, sites describe planning assumptions and cost estimates for long-term care in light of the anticipated end state of the site. The EM program will prepare a separate Stewardship Report that will discuss post-EM closure activities in more detail. Exhibit 2-7 presents the cumulative annual completion schedule for the EM sites. As shown in Exhibit 2-7, EM completed cleanup at 50 sites before 1997.

Exhibit 2-7
Site Completion Schedule

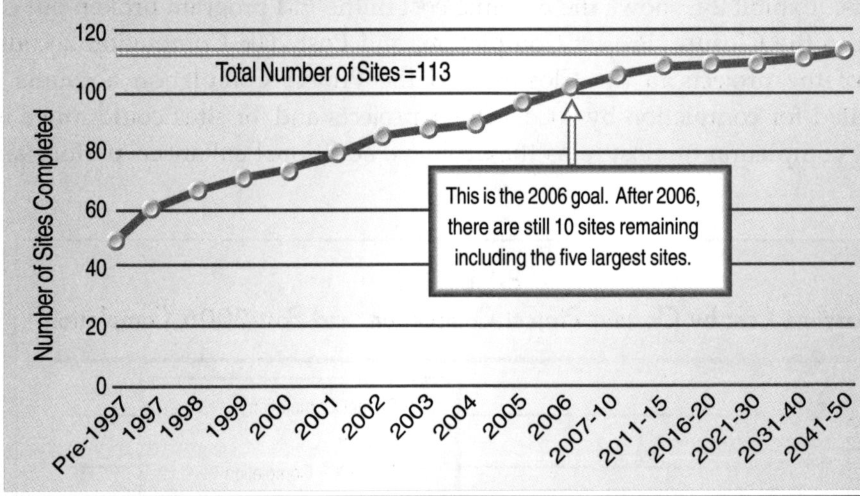

Exhibit 2-8 shows the planned baseline completion date for each site which had cleanup activities underway at the beginning of FY 1997. The exhibit is organized by state. Including sites completed prior to 1997, the EM program is estimating completion of 103 of 113, or over 90 percent, of the sites by 2006 for which the Environmental Management program had or has cleanup responsibility. This goal assumes that EM completes the Rocky Flats Environmental Technology Site and the Fernald Environmental Management Project by 2006 and 2005, respectively. If these goals are realized, only 10 sites will not complete their EM missions by 2006. Appendix C presents a complete list of all geographic sites.

Exhibit 2-8
Baseline Life-cycle Costs and Completion Dates By State

State	Operations/ Field Office	Site	Life-cycle Cost (in millions of constant 1998 dollars)[a]	Completion Date
Alaska	Nevada	Amchitka Island	7	2001
California	Albuquerque	Sandia National Laboratories - California	Included in SNL - NM	1999
California	Oakland	Energy Technology Engineering Center (ETEC)	229	2006
California	Oakland	General Atomics Site	11	2000
California	Oakland	General Electric Vallecitos Nuclear Center	21	2005
California	Oakland	Geothermal Test Facility	1	1997
California	Oakland	Laboratory for Energy-Related Health Research	22	2002
California	Oakland	Lawrence Berkeley National Laboratory	79	2003
California	Oakland	Lawrence Livermore National Laboratory Main Site	283	2006
California	Oakland	Lawrence Livermore National Laboratory Site 300	119	2006
California	Oakland	Stanford Linear Accelerator Center	5	2000
Colorado	Albuquerque	Grand Junction Office Site	15	2002
Colorado	Albuquerque	Maybell UMTRA Site	35	1998
Colorado	Albuquerque	Naturita UMTRA Site	60	1998
Colorado	Albuquerque	New Rifle UMTRA Site	9	1997
Colorado	Albuquerque	Old Rifle UMTRA Site	9	1997
Colorado	Albuquerque	Slick Rock Old North Continent UMTRA Site	4	1997
Colorado	Albuquerque	Slick Rock Union Carbide UMTRA Site	4	1997
Colorado	Nevada	Rio Blanco	12	2005
Colorado	Nevada	Rulison	4	1998
Colorado	Rocky Flats	Rocky Flats Environmental Technology Site	6,308	2010/ 2006[b]
Florida	Albuquerque	Pinellas Plant	263	1997
Idaho	Chicago	Argonne National Laboratory - West	14	2000
Idaho	Idaho	Idaho National Engineering and Environmental Laboratory	16,345	2050
Illinois	Chicago	Argonne National Laboratory - East	84	2002
Illinois	Chicago	Fermi National Accelerator Laboratory	2	1997
Illinois	Chicago	Site A	<1	1997
Iowa	Chicago	Ames Laboratory	1	1999
Kentucky	Albuquerque	Maxey Flats Disposal Site	13	2002
Kentucky	Oak Ridge	Paducah Gaseous Diffusion Plant	902	2010
Massachusetts	Oak Ridge	Ventron (FUSRAP Site)	NA	1997
Mississippi	Nevada	Salmon Site	9	1999
Missouri	Albuquerque	Kansas City Plant	83	1999
Missouri	Oak Ridge	Weldon Spring Site	365	2002
Nevada	Nevada	Central Nevada Test Site	19	2006
Nevada	Nevada	Nevada Test Site	2,149	2014
Nevada	Nevada	Shoal Site	18	2004

Exhibit 2-8 (Continued)
Baseline Life-cycle Costs and Completion Dates By State

State	Operations/ Field Office	Site	Life-cycle Cost (in millions of constant 1998 dollars)[a]	Completion Date
Nevada	Nevada	Tonopah Test Range Area	Included in Nevada Test Site	2007
New Jersey	Chicago	Princeton Plasma Physics Laboratory	11	1999
New Jersey	Oak Ridge	New Brunswick Site (FUSRAP Site)	NA	1997
New Mexico	Albuquerque	Los Alamos National Laboratory	1,578	2017
New Mexico	Albuquerque	Lovelace Respiratory Research Institute (formerly ITRI)	17	2000
New Mexico	Albuquerque	Sandia National Laboratories - NM	141	2001
New Mexico	Carlsbad	Waste Isolation Pilot Plant	7,722	2038
New Mexico	Nevada	Gasbuggy	10	2005
New Mexico	Nevada	Gnome-Coach	11	2004
New York	Chicago	Brookhaven National Laboratory	210	2006
New York	Oakland	Separations Process Research Unit (SPRU)	183	2014
New York	Ohio	West Valley Demonstration Project	1,114	2005
North Dakota	Albuquerque	Belfield UMTRA Site	0	1998
North Dakota	Albuquerque	Bowman UMTRA Site	0	1998
Ohio	Oak Ridge	Portsmouth Gaseous Diffusion Plant	835	2005
Ohio	Ohio	Ashtabula Environmental Management Project	93	2003
Ohio	Ohio	Columbus Environmental Management Project - King Avenue	22	1998
Ohio	Ohio	Columbus Environmental Management Project - West Jefferson	117	2005
Ohio	Ohio	Fernald Environmental Management Project	2,689	2008/ 2005[c]
Ohio	Ohio	Miamisburg Environmental Management Project	799	2005[d]
South Carolina	Savannah River	Savannah River Site	29,695	2038
Tennessee	Oak Ridge	Oak Ridge Reservation (including Y-12, ORNL, ETTP)	10,976	2013
Texas	Albuquerque	Pantex Plant	112	2002
Utah	Albuquerque	Monticello Millsite and Vicinity Properties	129	2001
Washington	Richland	Hanford Site	50,376	2046
Multiple States	NA	Long Term S&M Operations Office Costs Allocated to Multiple States	2,260	NA
Multiple States	NA	Program Direction Costs (Federal Salaries, Federal Travel, and Other Costs)	7,608	NA
Multiple States	NA	Technology Development Programs	2,885	NA
Multiple States	NA	All Other (Includes HQ and Other National Programs Costs)	143	NA

[a]Individual costs may not sum to $147.3 billion due to rounding.
[b]The Rocky Flats Environmental Technology Site is committed to accelerate activities to complete the site in 2006.
[c]The Ohio Field Office and the Fernald Environmental Management Project are committed to accomplishing completion scheduled for 2008 by the end of 2005.
[d]Pending validation of the current baseline, it is the goal of the Miamisburg Environmental Management Project and the Ohio Field Office to clean up the site by the end of 2003.

2.5 Maintaining Schedules

> **Programmatic Risk**
>
> Programmatic risk is defined as the risk to cost, schedule, and technical performance posed when an activity is not completed as scheduled. Sites document programmatic risk for activities on the critical closure path diagrams and on disposition maps. There are three categories of programmatic risk:
>
> - Technology (do we have the technology to do our work?)
> - Scope (do we know how much work there is to do?)
> - Intersite Dependency (do we know how and where we plan to store, treat, and dispose of material and waste?)

The EM program developed schedule estimates, making certain assumptions about the availability of funding. While the availability of funding is a critical influence on schedule, funding alone is not sufficient to ensure the successful completion of the objectives outlined in this document, which is based on numerous assumptions about scope and the achievement of key interim milestones.

To elevate key issues and focus management attention, sites have identified those activities and events (key interim milestones) that must occur if the EM program is to remain on schedule and correspondingly within cost. For these activities and events, sites have assigned a programmatic "risk" score in each of three areas: technology (do we have the technology to do our work?), scope (do we know how much work there is to do?), and intersite dependency (do we know how and where we plan to store, treat, and dispose of material and waste?). One example of such an activity is the signing of a Comprehensive Environmental Response, Compensation, and Liability Act (CERCLA) Record of Decision (ROD), through a process that must conform to regulatory requirements. In addition, some activities, such as the vitrification of high-level waste at the Hanford Site, can be completed only as quickly as capacity allows. In total, approximately 500 critical events and activities were reported for all sites. Exhibit 2-9 shows the distribution of programmatic risk scores among the three areas. Appendix D presents a detailed discussion of programmatic risk.

Exhibit 2-9
Distribution of Programmatic Risk Scores

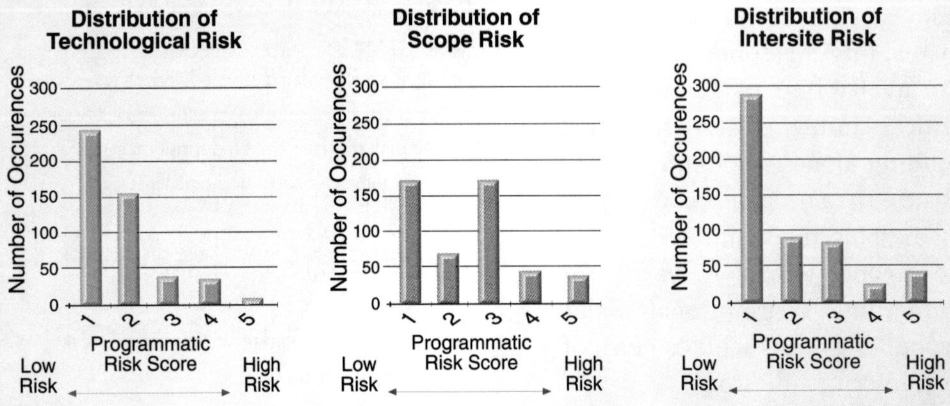

Sites identified more than 100 activities and events that had high programmatic risk scores (four or five on a scale of one to five) in any one of the three programmatic risk areas. Many of the activities that have a high programmatic risk score are crucial to the mission of the EM program. A high programmatic risk score means that the EM program must work diligently to ensure that those activities and events do not cause disruptions in schedule and subsequent increases in cost. One way EM is working to reduce programmatic risk is by ensuring that planned investments in science and technology are focused on the

Sample Critical Events and Activities

FY 1998, FY 1999, and FY 2000

- The Waste Isolation Pilot Plant opens for acceptance of transuranic waste in FY 1998.
- Nuclear material at the Fernald Environmental Management Project is packaged and shipped off site by September 1999.
- Fuel removal starts at the K-Basin at Hanford by July 1999.
- Records of Decision are signed at Oak Ridge for the East Tennessee Technology Park, Bethel Valley, Melton Valley, and Upper East Fork Poplar Creek between now and February 2000.
- West Valley selects a high-level waste receiving site by September 1998.
- The Savannah River Site is available to receive fluoride residues from the Rocky Flats Environmental Technology Site by April 1999 for stabilization.

critical events and activities with the highest technological risk. The text box lists a few of the high programmatic risk activities that must take place over the next three years. Critical activities and events that have high programmatic risk are discussed in the Operations/Field Office summaries in Chapter 3 and Appendix E.

2.6 Reconciliation with DOE FY 1997 Financial Statement

There are differences between the total life-cycle costs reported in *Paths to Closure* and the amount of unfunded environmental liabilities in the Department's FY 1997 financial statement. This section discusses the development of DOE's annual financial statement including the role of *Paths to Closure* and provides a reconciliation of the cost differences between the two documents.

The Government Management Reform Act of 1994 requires the Department of Energy to prepare annual audited financial statements reflecting the overall financial position of the Department, including assets and liabilities. The Act required submittal of the first financial statement by March 1, 1997 for the preceding fiscal year (FY 1996) and, for each year afterwards, requires the submittal of a statement by March 1 for the preceding fiscal year. By a significant margin, the Department's largest liability is its environmental liability.

The *Discussion Draft* is the basis for most of the environmental liability estimate in the Department's FY 1997 financial statement. The *Discussion Draft*, issued in June 1997, evolved into this report. Future DOE financial statements will rely on subsequent versions of *Paths to Closure* to estimate EM's portions of the Department's environmental liability. As a result of government-wide accounting principles to which federal government financial statements must conform and other reasons, there are differences between the FY 1997 DOE financial statement estimate of environmental liability and *Paths to Closure*. This section provides a reconciliation of the differences between the FY 1997 DOE financial statement and *Paths to Closure*.

The Department's FY 1997 Consolidated Statements of Financial Position[5] (financial statement) contains an unfunded environmental liability amount different from the EM cleanup life-cycle cost estimate in *Paths to Closure* for three reasons:

(1) The financial statement used the *Discussion Draft* as a basis for the EM life-cycle estimate due to the timing of financial statement publication;

(2) The financial statement makes adjustments to the EM estimate; and

(3) DOE has unfunded environmental liabilities in addition to the Environmental Management cleanup program described in *Paths to Closure*.

---

[5] As contained in *U.S. Department of Energy Fiscal Year 1997 Annual Report*, (DOE/CR-0057), Washington, DC, March 1998.

Exhibit 2-10 and the discussion that follows present a more detailed reconciliation between the *Paths to Closure* and the Department's FY 1997 financial statement estimates. As described in Chapter 1, there are several key differences between the *Discussion Draft* and *Paths to Closure*. The *Discussion Draft* contained a range of costs whereas *Paths to Closure* is a point estimate. The FY 1997 financial statement used the midpoint between the *Discussion Draft's* low and high planning scenarios (without enhanced performance).

Exhibit 2-10

Reconciliation Between *Paths to Closure* Life-cycle Cost Estimate and DOE FY 1997 Financial Statement Unfunded Environmental Liabilities

Line No.	Cost Element	Amount[a]	Comment
1	EM cleanup program (billions of 1998 dollars)	$147.3	Amount is total *Paths to Closure* life-cycle cost estimate.
2	Adjustments to reach EM cleanup program amount in financial statement including amount funded by current appropriations	(7.1)	Accounts for differences between *Paths to Closure* and *Discussion Draft* (used as basis for financial statement), conversion to 1997 dollars, and FY 1997 costs already incurred.
3	Active facilities	20.7	DOE estimate for deactivation and decommissioning of non-EM active facilities.
4	Pipeline facilities	8.7	DOE estimate for deactivation and decommissioning of non-EM inactive facilities from 1996 *Baseline Environmental Management Report* (BEMR).
5	High-level waste and spent nuclear fuel disposal	6.8	Represents DOE proportional share of Yucca Mountain repository life-cycle costs.
6	Other unfunded environmental liabilities	3.1	Represents $2.2 billion for excess plutonium dispositioning and about $0.9 billion for decontamination and decommissioning of inactive naval reactor facilities.
7	Total DOE unfunded environmental liabilities	179.5	Equals amount in the FY 1997 financial statement.

[a] All amounts are in billions of constant FY 1997 dollars to be consistent with the DOE FY 1997 financial statement, unless otherwise noted.

The DOE FY 1997 financial statement contains two adjustments to conform to government-wide accounting principles. First, because the financial statement is reported in constant 1997 dollars, it converts constant 1998 dollars. Second, the financial statement deducts funds spent during FY 1997.

The Department's FY 1997 financial statement contains four additional categories of unfunded DOE environmental liabilities beyond the Environmental Management cleanup program liabilities:

- **Deactivation and decommissioning of active facilities managed by DOE programs other than EM** (Line 3 of Exhibit 2-10). The Department estimates this category of environmental liability using EM deactivation and decommissioning models and information from the Department's corporate real property database, the Facilities Information Management System (FIMS).

- **Deactivation and decommissioning of surplus "pipeline" facilities not managed by EM but which are generally excess to the current mission of their programmatic owners** (Line 4 of Exhibit 2-10). Although not under EM management, these facilities were assumed to be candidates for transfer to the EM work scope. The 1996 *Baseline Environmental Management Report* (BEMR) chose to include these costs. Such costs will be included, in future *Paths to Closure* reports, after a decision is made to transfer the facilities to EM.

- **High-level waste and spent nuclear fuel disposal** (Line 5 of Exhibit 2-10). This estimate represents the Department's proportional share of the geologic repository life-cycle costs.

- **Other unfunded environmental liabilities** (Line 6 of Exhibit 2-10), including dispositioning of excess plutonium under the control of the Office of Defense Programs and decontamination and decommissioning of inactive naval reactor facilities.

Section 5.1.3 describes the relationship between ongoing changes to baselines, the future annual updates to *Paths to Closure,* and DOE's future financial statements.

Chapter 3

Environmental Management Site Cleanup Summaries

The current scope of the Environmental Management cleanup mission is described in many documents and management tools. Each product provides a different degree of detail and integration ranging from this document, *Paths to Closure*, that presents a national compilation of the cost, scope, and schedule challenges associated with the EM cleanup mission to the 353 individual Project Baseline Summaries (PBSs) that present the cost, scope, and schedule elements of each project. Exhibit 3-1 illustrates the relationship between these and other products.

Exhibit 3-1. *Paths to Closure* Documentation

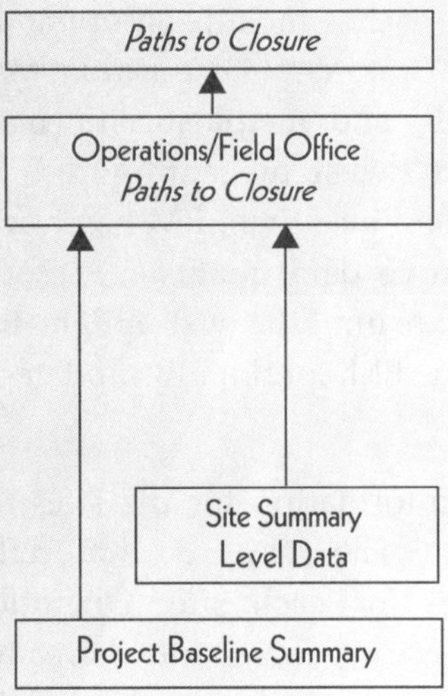

Single document providing integration and the national perspective.

- Albuquerque Operations Office
- Carlsbad Area Office
- Chicago Operations Office
- Idaho Operations Office
- Nevada Operations Office
- Oak Ridge Operations Office
- Oakland Operations Office
- Ohio Field Office
- Richland Operations Office
- Rocky Flats Field Office
- Savannah River Operations Office

Site-level descriptions of the current detailed cleanup intentions at the sites listed in Exhibit 2-7, including disposition maps and critical closure graphics.

353 Project Baseline Summary (PBS) documents providing technical details of individual projects. A complete file of PBSs is available for review on the EM homepage (http://www.em.doe.gov).

All of the documents and PBSs are further supported by site baselines and other detailed information maintained by the sites. This chapter and Appendix E present summaries of each Operations/Field Office's environmental management strategy. This chapter presents summaries of the Rocky Flats Field Office, the Richland Operations Office, and the Savannah River Operations Office. The summary of the Rocky Flats Field Office is described here because it illustrates a near-term closure effort with a challenging critical closure path. Rocky Flats must achieve significant enhanced performance goals if the site is to achieve the goal of closure by 2006. The Richland and Savannah River summaries are shown here because they illustrate the complexity of the cleanup effort associated with two other major DOE sites. Appendix E presents the EM cleanup summaries of the other eight Operations/Field Offices. The selection of Rocky Flats, Savannah River, and Richland as examples for Chapter 3 does not imply any priority between these sites and the others discussed in Appendix E.

The Rocky Flats Field Office, the Richland Operations Office, and the Savannah River Operations Office summaries that follow contain a discussion of the EM mission managed by the Operations/Field Office. The discussion is broken into five sections: a general overview; a discussion of end state assumptions; the cost and completion dates for the sites and projects; a work scope summary; and the critical closure paths and programmatic risks of the strategy managed under the Operations/Field Office. Additional information on all of the Operations/Field Offices can be found in the site versions of *Paths to Closure*.

Included as part of each work scope summary is a "Conceptual Summary Disposition Map." These maps show a summary of each office's current conceptual life-cycle approaches for managing EM wastes, nuclear materials, and contaminated media — from their current status, through storage, treatment, and disposal — to achieve the assumed site end states described in the relevant site strategy. In some cases, these conceptual approaches include shipping and off-site treatment and disposal. The Conceptual Summary Disposition Maps represent a "roll-up" from site-, waste-, material-, and media-specific maps. Volumes are approximate and have been rounded to two significant figures. The maps represent data approved as of February 1998. Since then, EM has carried out an effort to reconcile discrepancies and improve data quality. Although these improvements will not appear in *Paths to Closure* until the next update, they are reflected in the current "working" data set that EM continually updates as sites make changes.

Conceptual Summary Disposition Maps compile information for the sites that report through the Operations or Field Offices. The maps do not reflect Headquarters-directed or national-level strategies for each site, Operations Office, or Field Office. Within each map, activities are organized into "streams," which are defined as groups of materials, media, or wastes having similar origins, management requirements, or barriers to disposition. The following seven waste, material, and media categories are depicted in the maps:

- High-level waste (HLW)
- Transuranic waste (TRU)
- Mixed low-level waste (MLLW)
- Low-level waste (LLW)
- Environmental restoration activities (ER)
- Spent nuclear fuel (SNF)
- Nuclear materials

As has always been the case for this planning effort (reflected in December 1996 and October 1997 guidance to sites) implementation of each element of the EM program is contingent upon the completion of whatever evaluation is required under the National Environmental Policy Act (NEPA), the Comprehensive Environmental Response, Compensation, and Liability Act (CERCLA), or other statutes.

Decisions that remain to be made include those resulting from two DOE Environmental Impact Statements (EISs). Decisions on disposition of certain nuclear materials will be made pursuant to the Department's *Management of Certain Plutonium Bearing Residues and Scrub Alloys at the Rocky Flats Environmental Technology Site Environmental Impact Statement.* Until these decisions are made, the Conceptual Summary Disposition Maps reflect the "to be decided" (or "TBD") status of those materials.

Decisions on five waste types have been or will be made pursuant to the Department's May 1997 *Final Waste Management Programmatic Environmental Impact Statement* (WM PEIS). This nationwide NEPA analysis examined the potential environmental impacts of managing more than 2 million cubic meters of wastes from past, present, and future DOE activities. The Final WM PEIS identified preferred alternatives for transuranic waste treatment and storage, high-level waste storage, and hazardous waste treatment. The Department has identified preferred management strategies for mixed low-level waste treatment and disposal and low-level waste treatment and disposal. Preferred sites for these management activities have not yet been identified. In this chapter, assumptions regarding low-level and mixed low-level wastes are subject to change based on future Records of Decision (RODs). The Department has committed to publicly identify its preferred sites at least 30 days prior to issuing any ROD for these two waste streams. As of February 1998, one ROD has been issued from the WM PEIS process for transuranic waste treatment and storage. The Conceptual Summary Disposition Maps show specific disposition of transuranic waste, consistent with this ROD.

The Conceptual Summary Disposition Maps' depiction of environmental restoration activities differ from other waste or material management activities. Disposition paths for environmental restoration activities begin with "Contaminated Media" and show a "Response Strategy" for the media. Those strategies may or may not be based on decisions regarding environmental restoration wastes resulting from the CERCLA, NEPA, and Resource Conservation and Recovery Act (RCRA) processes. Where such decisions have not yet been made, environmental restoration planning was based upon assumptions that are being evaluated under CERCLA, NEPA, and/or RCRA, and may change as more media characterization data become available, as comments are received from local stakeholders through public involvement processes, or as the regulatory agencies review and evaluate the various cleanup alternatives.

3.1 Rocky Flats Field Office Summary

The Rocky Flats Environmental Technology Site (RFETS) is located approximately 15 miles northwest of Denver, Colorado. Construction of the site started in 1951. Facilities at the site are located on approximately 385 acres of an industrial area, surrounded by a buffer zone of approximately 5,800 acres of prairie terrain. RFETS has over 700 permanent structures that were built to support its mission. The primary mission of the site was the manufacture and assembly of nuclear and nonnuclear weapons components, as well as to recover plutonium. In January 1992, the nuclear weapons production mission of the site was terminated formally; the nonnuclear mission of the site was completed in October 1994. The only remaining mission of the site is cleanup and remediation. The potential risks to health and safety at RFETS arise principally from the large amounts of special nuclear materials (SNM), residues contaminated with plutonium, and radioactive wastes that are stored at the site.

Rocky Flats Environmental Technology Site

1998

End State

3.1.1 End State

Intermediate site condition expectations for RFETS were developed through a detailed discussion, negotiation, and approval process that resulted in the Rocky Flats Cleanup Agreement (RFCA). Approved in July 1996, this agreement establishes a legally binding relationship between the U.S. Department of Energy (DOE), the Environmental Protection Agency, and the Colorado Department of Public Health and Environment that governs cleanup at the site.

According to the RFCA, planned cleanup levels will permit open space use of the site's buffer zone, and the industrial area will be cleaned up for restricted open space or industrial reuse. Approximately 100 acres of the site will be capped where complete remediation is technically or economically infeasible. The caps will reduce water infiltration and direct runoff in the area, thereby preventing migration of contaminants. Additional cleanup may be conducted should technological advances or increased funding allow.

Post-closure stewardship requirements for the site have not yet been determined. DOE is currently participating in discussions with the community to determine when it will be appropriate to make long-term stewardship decisions and what the future use of the site should be. DOE expects that discussions about future use may continue for several years before community sentiment is well understood and the site is ready to investigate implementation. Additional information about the RFETS intermediate site condition and long-term stewardship can be found in the Rocky Flats version of *Paths to Closure.*

3.1.2 Cost and Completion Date

The Rocky Flats Field Office has separated its closure activities into 29 discrete projects. The Project Baseline Summary (PBS) developed for each project sets forth detailed strategies for completion of the project and programmatic information that includes cost, schedule, scope, end state, and interim milestones. Exhibit 3-2 presents a summary of the Rocky Flats cost and schedule information for these projects. Additional information is available in each PBS.

The estimated EM life-cycle cleanup cost for the Rocky Flats Environmental Technology Site is $6.3 billion (constant 1998 dollars). The Rocky Flats cost estimate includes several years of long-term surveillance and monitoring. These costs will be incurred after cleanup activities are completed. Given the uncertainty associated with outyear costs, specifically the cost and duration of stewardship activities, these costs will continue to be refined.

While the March 1997 baseline indicates that the site completion date for the RFETS is 2010, both EM Headquarters and the Rocky Flats Field Office have undertaken the challenge of completing all closure work by the year 2006. To accomplish that challenge, significant enhanced performance goals must be achieved. The management approach, scheduling impacts, technical development, and intersite integration needed to accomplish this goal of completion by 2006 are discussed in more detail in the Rocky Flats Field Office version of *Paths to Closure.* The Rocky Flats Field Office is in the process of revising the 2010 baseline to reflect the commitment to the 2006 goal. The documentation for a 2006 baseline will be completed by the end of this calendar year.

Exhibit 3-2 Rocky Flats Field Office
Cleanup Project Summary: Duration and Costs (All costs in thousands of 1998 dollars)[a]

Site Closure Project Activities	1997 - 2006	2007 - 2070	Total
Rocky Flats Field Office			
Work for Others Project	790	0	790
Rocky Flats Environmental Technology Site (RFETS)			
New Plutonium Interim Storage Vault	1,688	0	1,688
Uranium Disposition Project	12,984	0	12,984
Plutonium Liquid Stabilization	41,496	0	41,496
Building 779 Cluster Closure Project	21,093	0	21,093
Special Nuclear Materials Capital Support	17,409	0	17,409
Infrastructure Improvement/ Replacement Project	84,606	0	84,606
Plutonium Solid Residue Stabilization Project	454,286	0	454,286
Plutonium Metals and Oxides Stabilization	63,105	0	63,105
Special Nuclear Materials Shipping Project	25,308	0	25,308
International Atomic Energy Agency Project	1,174	0	1,174
Building 991 Cluster Closure Project	15,148	0	15,148
Building 771/774 Cluster Closure Project	189,446	906	190,352
Building 776/777 Cluster Closure Project	195,640	13,385	209,024

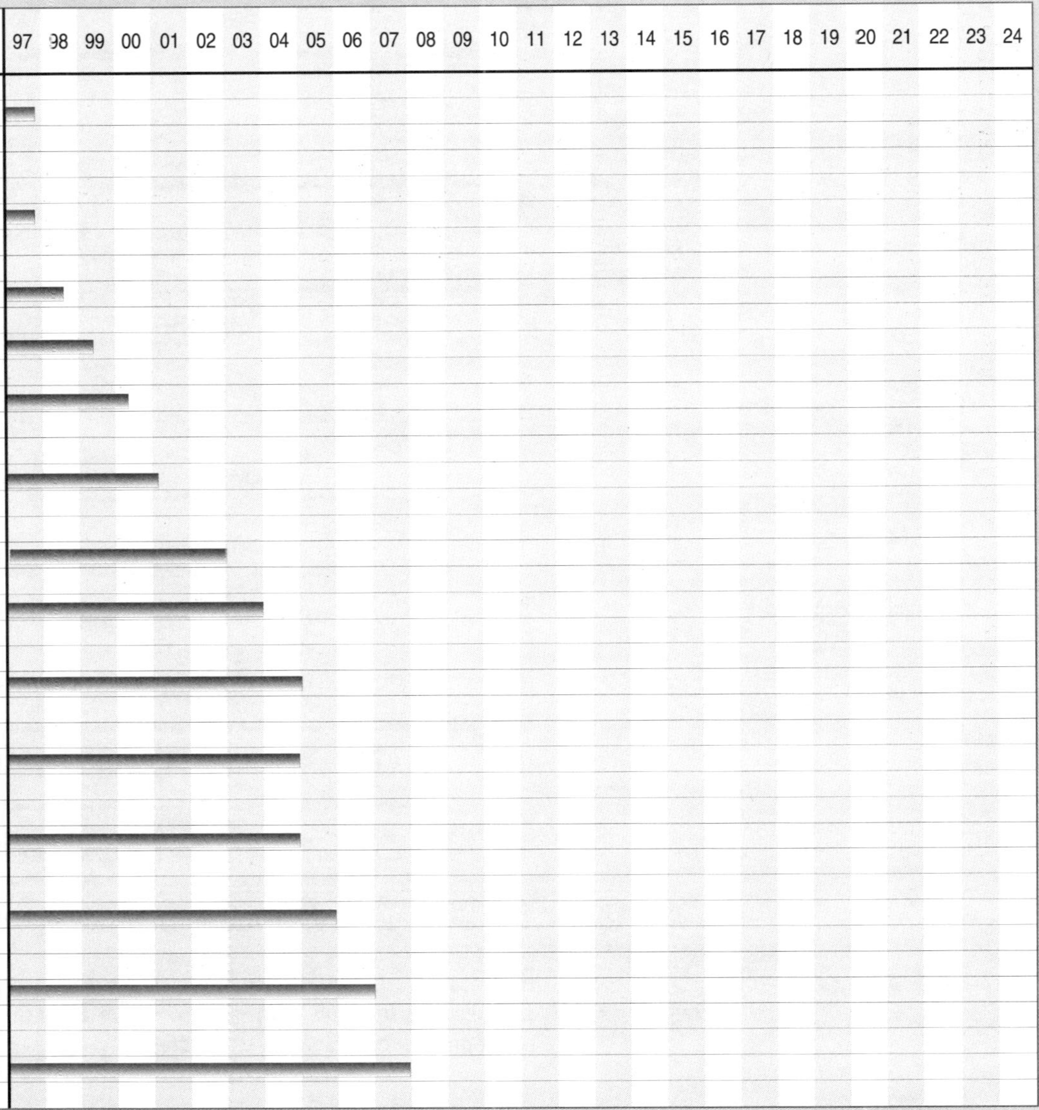

Exhibit 3-2 Rocky Flats Field Office (Continued)
Cleanup Project Summary: Duration and Costs (All costs in thousands of 1998 dollars)

Site Closure Project Activities	1997 - 2006	2007 - 2070	Total
Building 371 Cluster Closure Project	274,377	21,535	295,912
Building 707/750 Cluster Closure Project	209,292	8,880	218,172
Building 881 Cluster Closure Project	57,729	20,020	77,749
Industrial Zone Closure Project	232,453	56,848	289,301
Rocky Flats Field Office - DOE Management	214,349	304,300	518,649
Miscellaneous Production Zone Cluster Closure Project	99,448	21,237	120,686
Safeguards and Security Project	345,109	28,347	373,457
Buffer Zone Closure Project	171,644	46,767	218,411
Utilities & Infrastructure Project	550,263	75,298	625,561
Waste Management Project	844,250	161,190	1,005,440
Remediation Waste & Contingent Storage Project	5,189	3,691	8,880
Closure Caps Project	0	61,921	61,921
K-H Project Management	1,109,129	122,351	1,231,480
Analytical Services Project	59,457	14,644	74,101
Special Nuclear Materials Consolidation Project	49,597	0	49,597
Total	**5,346,459**	**961,320**	**6,307,779**

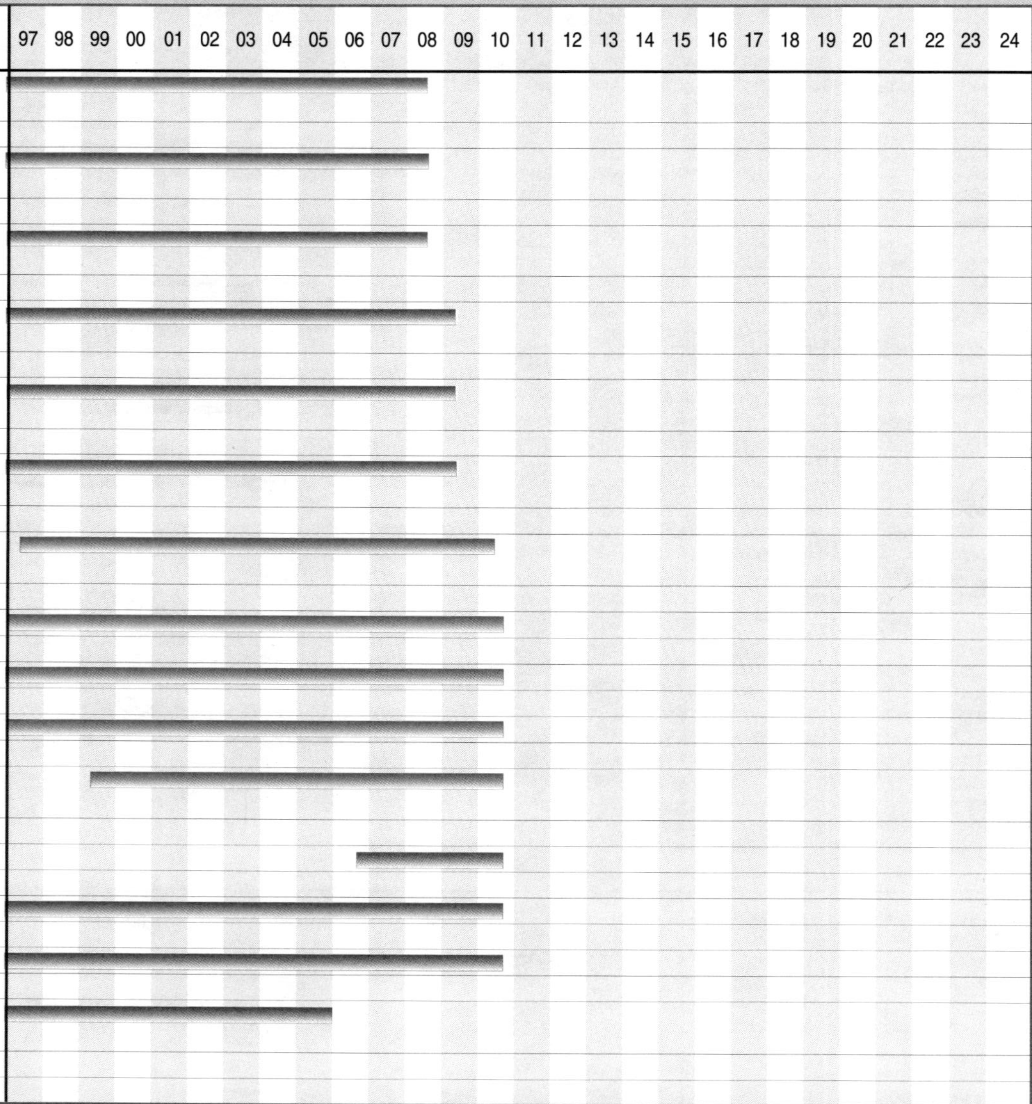

[a] The Rocky Flats Field Office is in the process of revising the 2010 baseline to reflect the commitment to the 2006 goal.

The projected cost profile associated with the closure of RFETS was developed by combining the cost estimates presented in each Project Baseline Summary. Exhibit 3-3 displays the resultant baseline cost profile.

Exhibit 3-3
Rocky Flats Field Office Environmental Management Baseline Cost Profile[a]

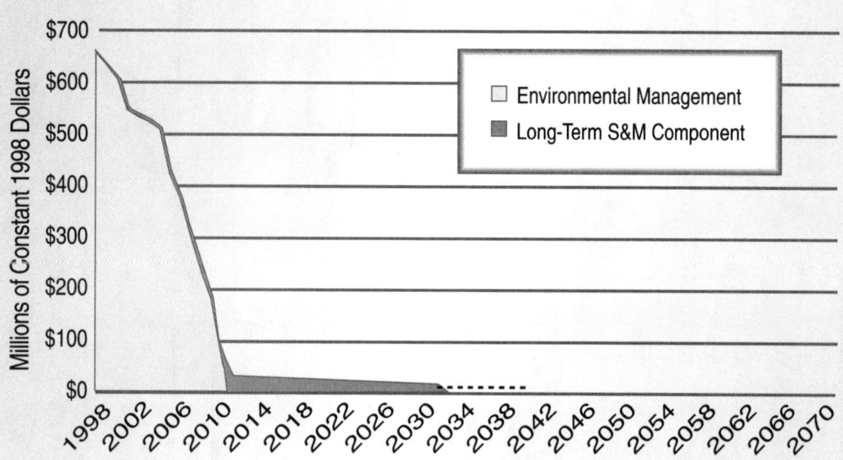

[a] Upon achievement of the 2006 closure goal, the only remaining cost would be for long-term surveillance and monitoring.

3.1.3 *Work Scope Summary*

The scope of work necessary to achieve closure as defined in the Rocky Flats Cleanup Agreement includes the stabilization and management of plutonium metals, oxides, residues, and solutions; enriched uranium metals and oxides; and wastes generated from closure activities. Existing waste and materials, as well as the waste generated from the cleanup, will be treated (if required), packaged, and transported according to off-site waste acceptance criteria and all applicable laws and regulations. The sections below describe the major waste, material, and contaminated media volumes to be addressed by the Rocky Flats Field Office. The volumes reported are approximate, and correspond to the major waste, material, and media flows, the potential treatment processes, and the off-site disposal destinations presented in Exhibit 3-4, the Rocky Flats Field Office Conceptual Summary Disposition Map.

Exhibit 3-4

Rocky Flats Field Office Conceptual Summary Disposition Map

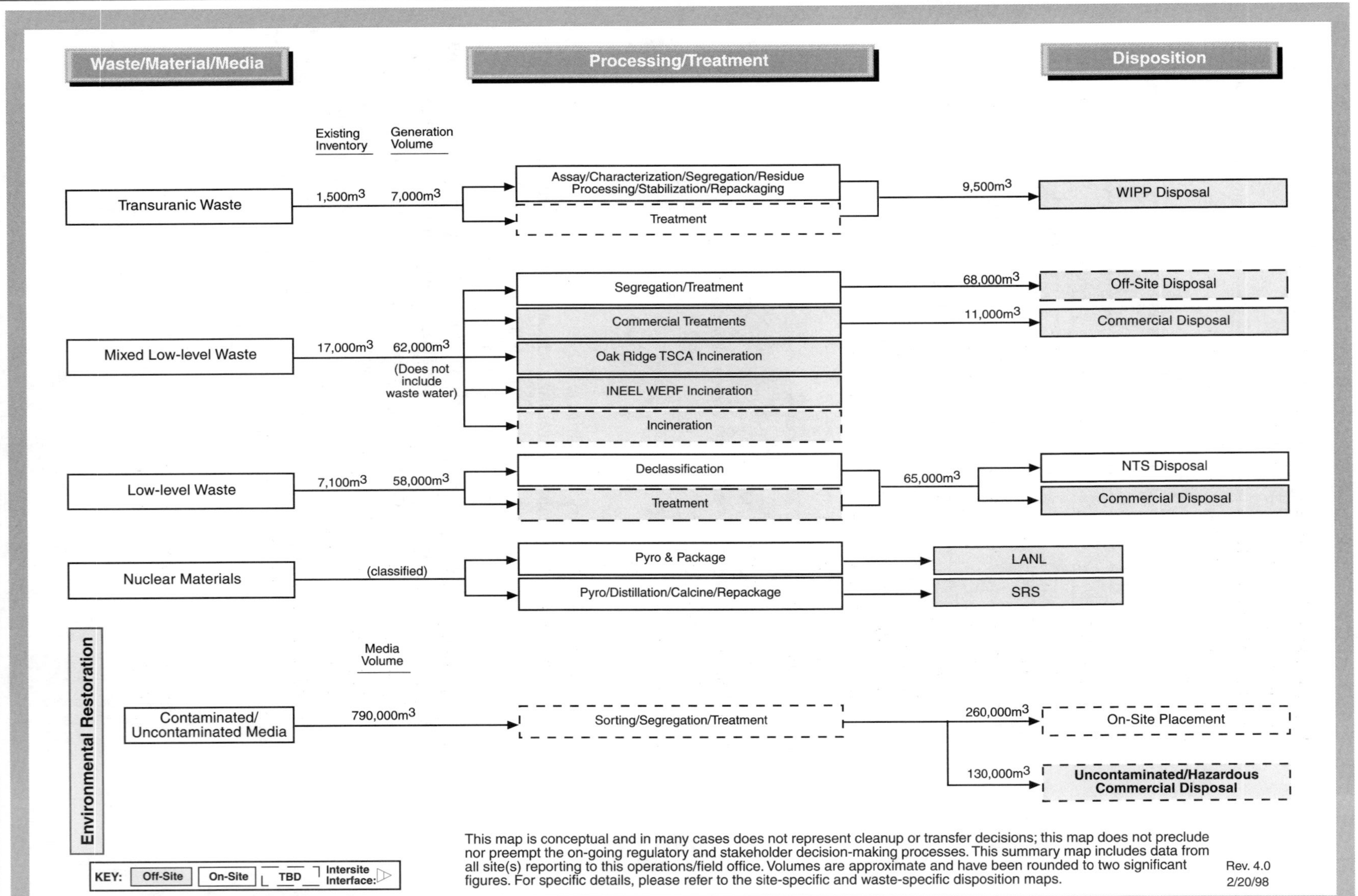

This map is conceptual and in many cases does not represent cleanup or transfer decisions; this map does not preclude nor preempt the on-going regulatory and stakeholder decision-making processes. This summary map includes data from all site(s) reporting to this operations/field office. Volumes are approximate and have been rounded to two significant figures. For specific details, please refer to the site-specific and waste-specific disposition maps.

Rev. 4.0
2/20/98

Transuranic Waste

- Approximately 1,500 cubic meters of legacy transuranic waste are currently in inventory and an estimated 7,000 cubic meters of transuranic waste are expected to be generated over the life cycle of operations. After treatment and repackaging, 9,500 cubic meters of transuranic waste are planned to be shipped to WIPP.

Other Waste

- Approximately 17,000 cubic meters of mixed low-level waste are currently in inventory (primarily "Pondcrete" and Solar Pond sludge) and 62,000 cubic meters of mixed low-level waste are estimated to be generated over the life cycle of operations (including waste generated by remedial action and facility deactivation and decommissioning). While decisions on the treatment and disposition of this material will be made in Records of Decision, resulting from CERCLA and the Waste Management Programmatic Environmental Impact Statement (WM PEIS), it is assumed that approximately 11,000 cubic meters may be treated and disposed of at an off-site commercial facility and an additional 68,000 cubic meters may be disposed of off site at a location to be determined later.

- Approximately 7,100 cubic meters of low-level waste are in inventory and 58,000 cubic meters of low-level waste are estimated to be generated over the life cycle of operations (including waste generated by remedial action and facility deactivation and decommissioning activities). While decisions on the treatment and disposition of this material will be made in Records of Decision resulting from CERCLA and the WM PEIS, it is assumed that after declassification and treatment of some low-level waste, 65,000 cubic meters may be disposed of at the Nevada Test Site and an off-site commercial facility.

Remedial Action and Facility D&D

- Approximately 790,000 cubic meters of environmental media (including 300,000 cubic meters of groundwater, 198,000 cubic meters of soils, and nearly 295,000 cubic meters of facility deactivation and decommissioning generated material) contaminated with radionuclides (including transuranic elements) and hazardous substances will be managed. After segregation and treatment, a total of 260,000 cubic meters are expected to be placed on site and 130,000 cubic meters are expected to be disposed of at an off-site commercial facility.

Nuclear Materials

- Nuclear materials volumes are classified and cannot be disclosed in this document.

The closure mission at RFETS is documented in projects that involve waste, special nuclear material (SNM), facility deactivation and decommissioning, and environmental restoration. Work in each of those areas is planned, funded, and executed under a comprehensive risk reduction strategy that places a priority on maintaining safety at the site, thereby ensuring the continued safety of site workers, the public, and the environment, and then eliminating the site's highest priority risks. Activities which address the site's highest priority risks, in order of priority are: stabilization, consolidation, and packaging of SNM; shipment of SNM; deactivation of nuclear facilities (to reduce facility baseline costs); waste management; and facility decommissioning and environmental restoration. Long-term groundwater treatment and surveillance and monitoring, the scope of which is yet to be determined, will continue after closure.

At RFETS, the bulk of costs are driven by continued storage of SNM, residues, and wastes. Each building closure and infrastructure project integrates all activities necessary to continue safe operations and to eliminate buildings, including operation and maintenance of safety envelopes, deactivation, decontamination (to the extent necessary), decommissioning, dismantlement, and environmental remediation of the land under the buildings. The remainder of the work scope includes environmental remediation of land areas outside building footprints, including the buffer zone. Groundwater will be passively remediated and post-closure environmental monitoring will be required after site closure. The scope of the post-closure requirements will be described in the CERCLA Record of Decision at closure. Exhibit 3-5 displays RFETS site closure costs by major work scope category.

Exhibit 3-5
Rocky Flats Field Office
Life-Cycle Costs by Category[a]

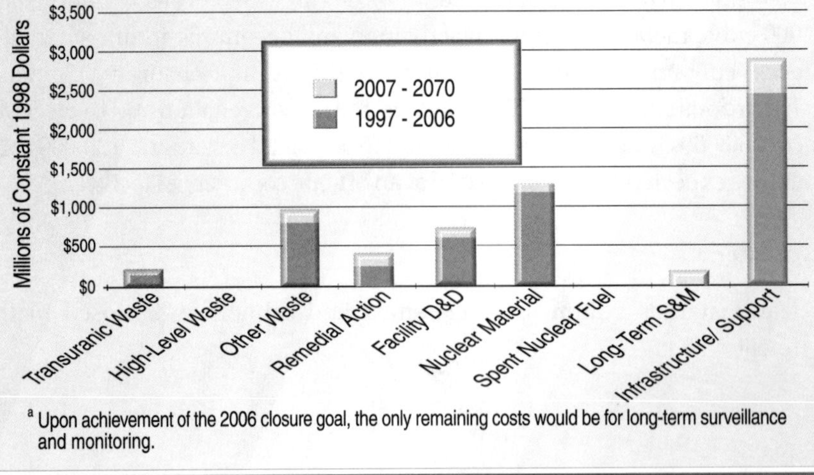

[a] Upon achievement of the 2006 closure goal, the only remaining costs would be for long-term surveillance and monitoring.

3.1.4 Critical Closure Path and Programmatic Risk

The critical closure path schedule presented in Exhibit 3-6 sets forth the timetable for completing the closure activities at RFETS. The highlighted activities show the critical closure path, which represents the series of events that drive the overall completion date for the site. In Exhibit 3-6, the bars represent projects and critical activities, and the triangles represent critical events and milestones.

The primary key for RFETS to close on schedule is the ability to ship materials and wastes to receiver sites. The site is consolidating nuclear materials into fewer buildings to minimize operations and costs and maximize the funding available for closure activities. However, the key activity on the critical closure path in the early years is the stabilization of nuclear materials and their packaging in configurations certified for shipping. RFETS has developed a closure project plan that minimizes the total project cost by balancing the nuclear materials preparation activities (risk reduction) with building elimination ("mortgage" reduction). In an effort to further accelerate the closure schedule, activities that have the potential to improve the efficiency of those two efforts are being identified and evaluated for implementation.

Completion of the EM mission at the Rocky Flats Field Office as scheduled will depend on the timely accomplishment of critical activities and events, some of which are external milestones (external milestones are those that are beyond the ability of the site to resolve). Exhibit 3-7 presents a summary of activities/ milestones on the critical closure path that have high programmatic risk (programmatic risk scores of 4 or 5 in any category). In addition to those high programmatic risk milestones, several other external milestones have an effect on the site's ability to achieve its closure goal. Those milestones include the ability of potential receiver sites to receive materials from the Rocky Flats Environmental Technology Site and the availability of safe, secure transport of the materials to receiver sites.

This page intentionally left blank.

Exhibit 3-6

Rocky Flats Field Office Critical Closure Path

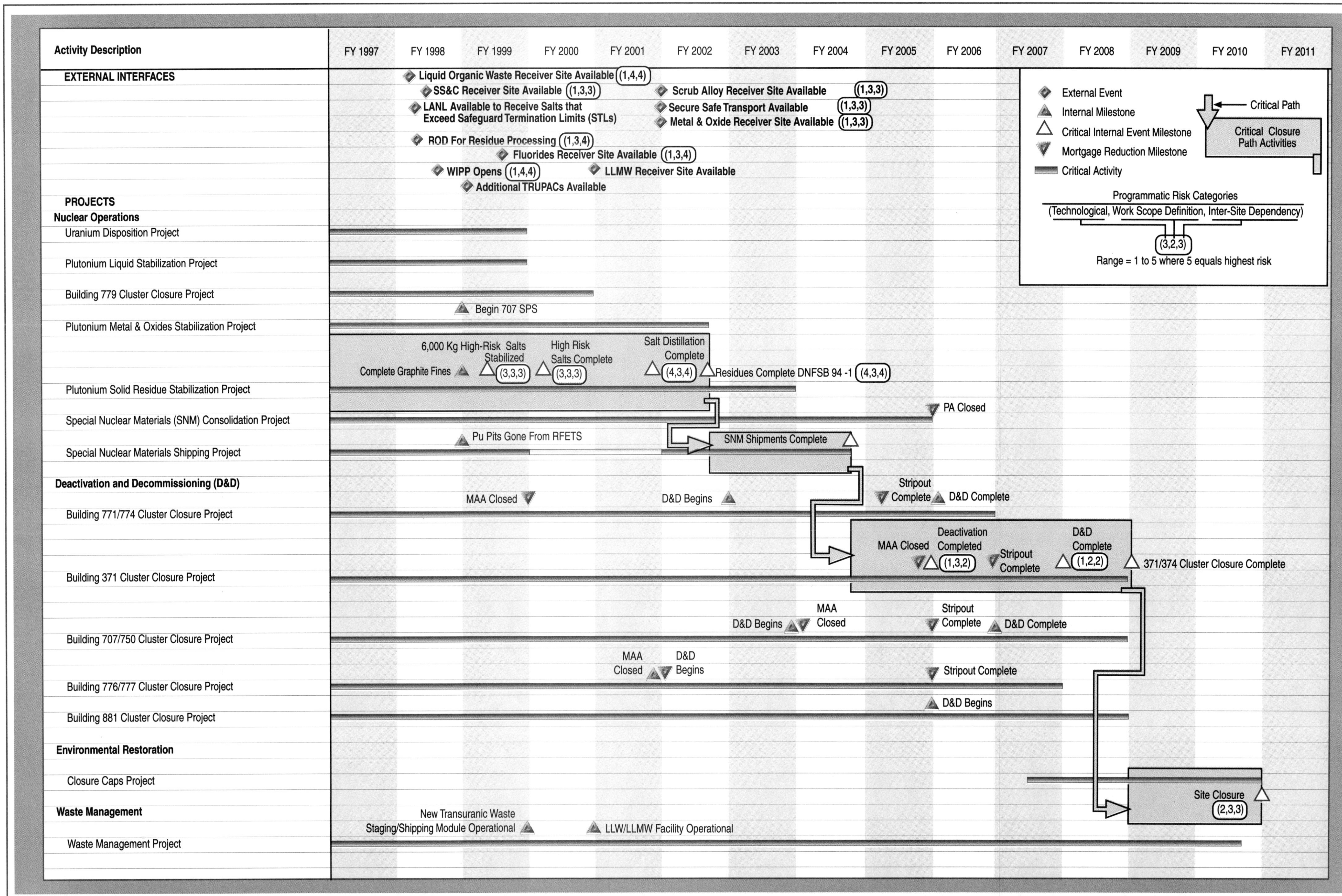

Exhibit 3-7

Summary of High Programmatic Risk Activities/Milestones:

Rocky Flats Field Office

Project, Activity, Event	Start/ End Date	Programmatic Risk Categories		
		Technological	Work Scope Definition	Intersite Dependency
ORNL available to receive organic waste liquid for treatment and disposition	Dec 97	1	4	4
HQ Residue Processing Record of Decision issued	May 98	1	3	4
WIPP opens for receipt of RFETS TRU waste	May 98	1	4	4
SRS available to receive fluoride residues for stabilization	Apr 99	1	3	4
Salt distillation complete	2001	4	3	4
Complete stabilization of all solid residues (Complete DNFSB 94-1 commitments)	May 02	4	3	4

3.2 Richland Operations Office Summary

The Richland Operations Office manages the cleanup work at the Hanford Site. The Hanford Site occupies 560 square miles in southeastern Washington State. It was acquired by the federal government in 1943 for the first full-sized plutonium production operation. The Hanford Site has been used for a variety of purposes, including plutonium production, chemical processing, waste management, and research and development activities.

Richland Operations Office (Hanford Site)

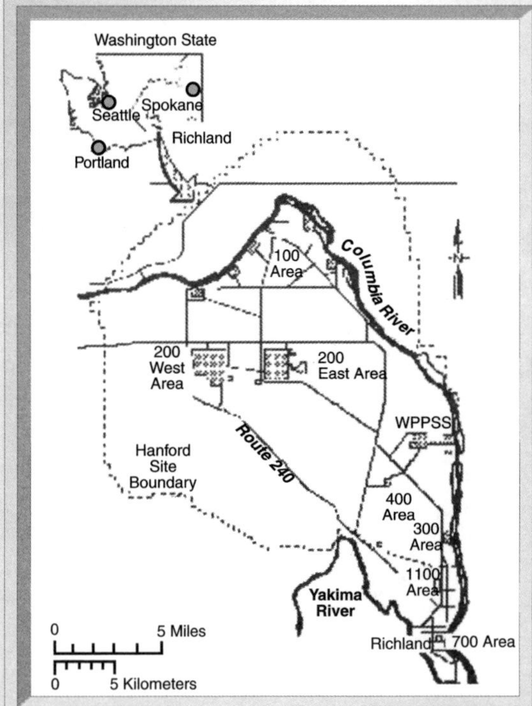

The current mission of the Hanford Site is to manage the facilities and inventories of special materials, remedy the environmental contamination caused by decades of activities related to the production of plutonium, and support national research efforts in the areas of environmental cleanup and other sciences. The major Hanford Site cleanup mission areas include the Tank Waste Remediation System (TWRS) project, the Waste Management project, the Facility Transition project, the Environmental Restoration project, the Science and Technology project, and other supporting projects.

After the defined Environmental Management cleanup mission is completed at the Hanford Site, the federal government will continue in a caretaker role due to disposed waste remaining on site. Ongoing missions at the Hanford Site will also continue primarily in the areas of science and technology development.

3.2.1 End State

Alternatives for potential future use of the Hanford Site lands were developed through a cooperative effort with the U.S. Department of Energy (DOE); the Confederated Tribes of the Umatilla Indian Reservation; the Nez Perce Tribe; the United States Department of the Interior; the City of Richland; and Benton, Franklin, and Grant Counties. These alternatives are being analyzed in the Hanford Remedial Action Environmental Impact Statement (HRA-EIS) and Comprehensive Land Use Plan for the potential environmental impacts resulting from the proposed future land uses associated with each alternative. As mandated by

Public Law 104-201, Section 3153, the land-use plan will address a 50-year planning period. Once established, the land-use plan will provide a framework for making land-use and facility-use decisions while DOE manages the land.

The selection of the appropriate land uses for the Hanford Site will be made following the decision-making processes described earlier in Section 1.3. When sites are certified as complete, any CERCLA and RCRA requirements for long-term surveillance, monitoring, and maintenance will be identified along with the appropriate institutional controls to protect human health and the environment. The planning end state of the Hanford Site will be developed in the Comprehensive Land Use Plan.

Currently, the assumption is that the federal government will remain the landlord of the site after cleanup is complete. Cleanup levels and disposal standards will be established that are consistent with projected long-term uses; and remediation will be performed to ensure the protection of human health, the environment, and the Columbia River. Groundwater use remains restricted indefinitely.

The 100 Area of the site lies along the Columbia River and is comprised of over 400 waste sites, nine retired plutonium production reactors, and their ancillary facilities. Residential cleanup standards have been established for remediation in the area. The C-Reactor was placed into Interim Safe Storage, with plans to place seven of the other reactors into safe storage. The B-Reactor structure is expected to remain as a National Historic Landmark. Groundwater remediation is being performed to protect the Columbia River.

The 200 Area of the site is expected to be maintained as a waste management area. Waste from on-site and off-site sources will be stored and disposed in the 200 Area. The Environmental Restoration Disposal Facility (ERDF) will accept waste that meets acceptance criteria from all Hanford CERCLA sites, and will be expanded to have a capacity of more than 4 million cubic yards of waste. Approximately 700 waste sites will be remediated in the 200 Area. Remediation is expected to be completed through a combination of waste excavation and placement of soil barriers over waste sites. Tank waste will be retrieved and immobilized from the 177 high-level waste tanks. The low-level waste burial grounds will be stabilized and the RCRA storage facilities will be RCRA clean-closed unless required for the ensuing caretaker mission.

The 300 Area is being remediated to meet industrial cleanup standards. Soil remediation is being performed to remediate over 100 waste sites. Facilities which will not be turned over to the private sector for further use will be demolished.

Though final end states have not been set for the site, it is anticipated that the land near the Columbia River would be remediated for recreational use. Additional information about Richland end states and long-term stewardship can be found in the Richland Operations Office version of *Paths to Closure*.

3.2.2 Cost And Completion Dates

The Richland Operations Office has divided its environmental management work into 45 discrete projects. A Project Baseline Summary (PBS) exists for each project and contains detailed programmatic information, including cost, schedule, scope, end state, and interim milestones. The projected cost profile associated with the Richland Operations Office is developed by combining the cost estimates from each PBS. Exhibit 3-8 displays the resultant baseline cost profile. A summary of the cost and schedule information for each project is illustrated in Exhibit 3-9. For additional information about these projects, see each PBS.

Exhibit 3-8
Richland Operations Office
Environmental Management Baseline Cost Profile[a]

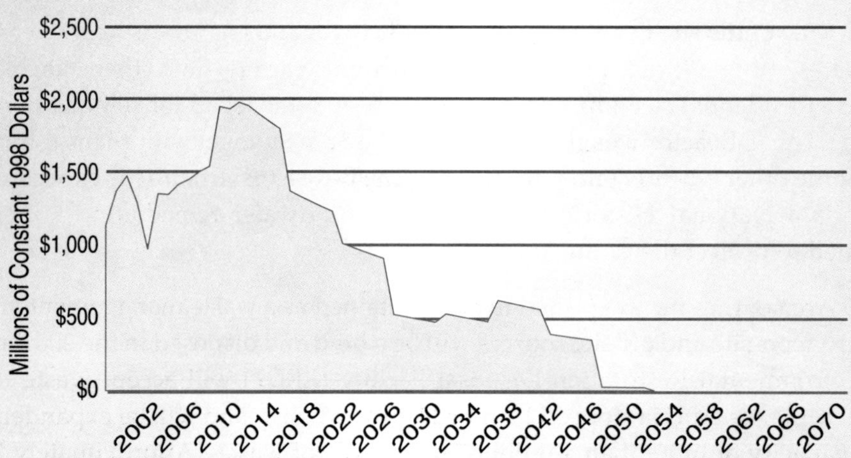

[a]Does not include long-term surveillance and monitoring cost.

The estimated life-cycle cost for cleanup of the Hanford site is $50.3 billion (constant 1998 dollars). This estimate does not include $500 million (constant 1998 dollars) in non-EM costs or the costs associated with federal oversight (i.e., program direction). This baseline cost profile does not reflect any potential effects of budgetary funding constraints which will likely affect the overall life-cycle cost of Hanford Site cleanup. The current baseline supports the completion of EM work (excluding long-term surveillance and monitoring) by 2046.

Exhibit 3-9 Richland Operations Office
Cleanup Project Summary: Duration and Costs (All costs in thousands of 1998 dollars)

Site Closure Project Activities	1997 - 2006	2007 - 2070	Total
Richland Operations Office			
TWRS Regulatory Unit	34,536	0	34,536
RL Directed Support	215,071	467,194	682,265
PNNL Waste Management	160,802	627,828	788,630
Hanford			
Environmental Restoration Disposal Facility	233,922	258,262	492,183
N Reactor Deactivation	28,220	0	28,220
PUREX Sub-Project	22,593	0	22,593
B-Plant Sub-Project	51,622	0	51,622
300 Area/SNM Sub-Project	20,181	0	20,181
Tank Waste Characterization	212,794	0	212,794
Advanced Reactors Transition	68,574	0	68,574
PFP Stabilization	159,985	0	159,985
Spent Nuclear Fuels Project	846,839	0	846,839
324/327 Facility Transition Project	192,482	0	192,482
TWRS Management Support	246,033	0	246,033
Tank Safety Issue Resolution Project	127,079	0	127,079
Tank Farms Operations	849,488	23,722	873,210
K Basin Deactivation	89,322	20,406	109,728

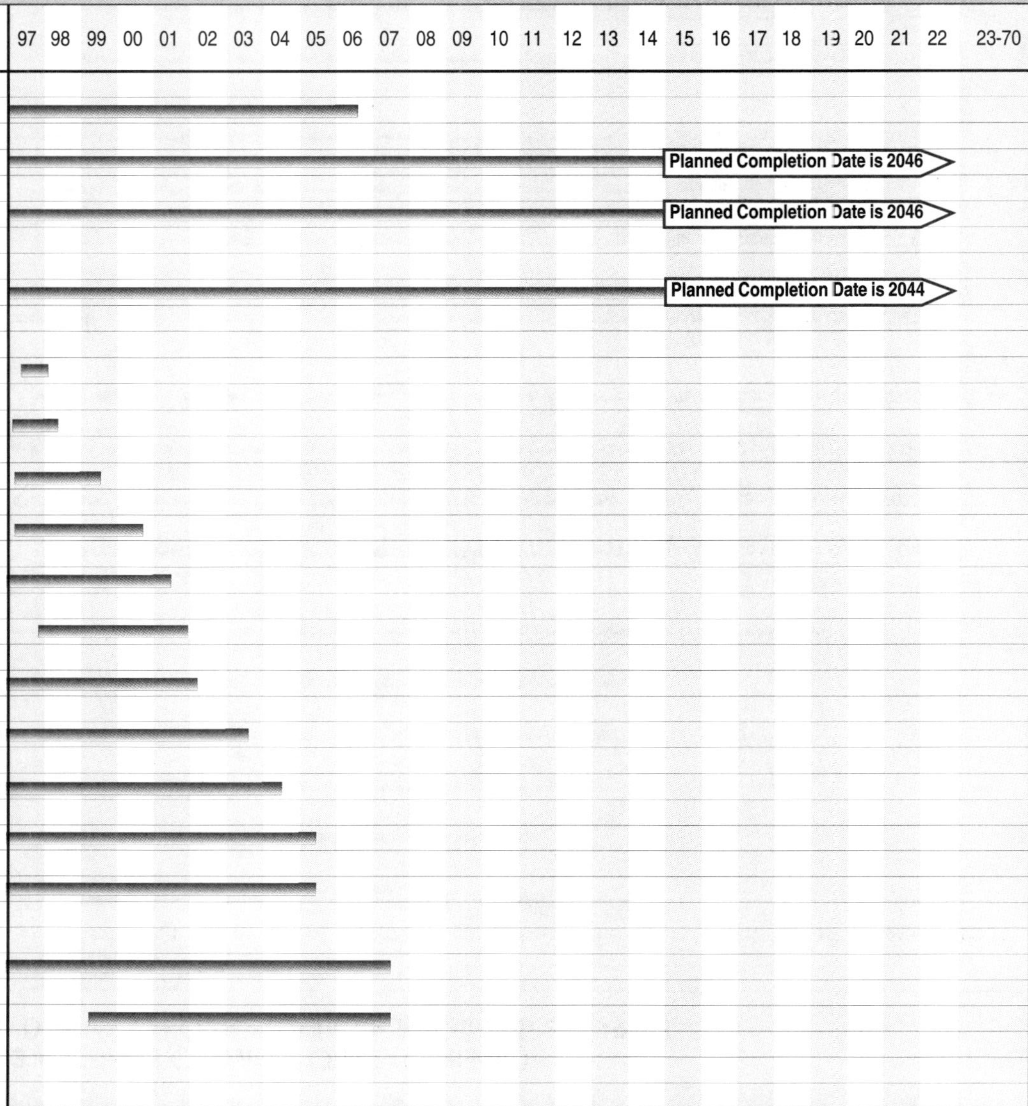

Exhibit 3-9 Richland Operations Office (Continued)
Cleanup Project Summary: Duration and Costs (All costs in thousands of 1998 dollars)

Site Closure Project Activities	1997 - 2006	2007 - 2070	Total	97 98 99 00 01 02 03 04 05 06 07 08 09 10 11 12 13 14 15 16 17 18 19 20 21 22 23-70
Hanford Surplus Facility Program 300 Area Revitalization Project	98,929	9,058	107,987	
100 Area Remedial Action	347,409	182,289	529,698	
Process Waste Privatization Phase I	2,940,124	1,785,788	4,725,911	
300 Area Remedial Action	17,565	136,528	154,093	
PFP Deactivation	68,341	83,521	151,862	
Waste Encapsulation and Storage Facility Sub-project	112,019	97,883	209,903	
200 Area Remedial Action	218,998	1,614,334	1,833,332	Planned Completion Date is 2026
PFP Vault Management	552,180	661,279	1,213,459	Planned Completion Date is 2028
Process Waste Support	127,410	660,356	787,766	Planned Completion Date is 2028
Decontamination and Decommissioning	144,697	1,596,107	1,740,803	Planned Completion Date is 2030
Process Waste Privatization Phase II	300,458	10,358,316	10,658,775	Planned Completion Date is 2031
Liquid Effluents Project	348,883	585,893	934,776	Planned Completion Date is 2032
Solid Waste Treatment	504,982	977,616	1,482,598	Planned Completion Date is 2035
Process Waste Privatization Infrastructure	272,700	1,486,773	1,759,473	Planned Completion Date is 2036
Transition Project Management	110,957	108,493	219,450	Planned Completion Date is 2037
Accelerated Deactivation	35,227	243,488	278,714	Planned Completion Date is 2037
Post Closure Surveillance & Maintenance	3,206	55,778	58,984	Planned Completion Date is 2043

Exhibit 3-9 Richland Operations Office (Continued)
Cleanup Project Summary: Duration and Costs (All costs in thousands of 1998 dollars)

Site Closure Project Activities	1997 - 2006	2007 - 2070	Total	97 98 99 00 01 02 03 04 05 06 07 08 09 10 11 12 13 14 15 16 17 18 19 20 21 22 23-70
Facility Surveillance & Maintenance - ADS 3500	128,502	302,618	431,119	Planned Completion Date is 2043
Groundwater Management	202,808	335,639	538,447	Planned Completion Date is 2043
Program Management and Support	431,430	1,252,843	1,684,273	Planned Completion Date is 2044
Canister Storage Building Operations	12,530	412,048	424,578	Planned Completion Date is 2046
Retrieval Project	983,359	2,985,590	3,986,949	Planned Completion Date is 2046
Immobilized Tank Waste Storage & Disposal Project	332,828	6,831,983	7,164,811	Planned Completion Date is 2046
Landlord Project	177,831	269,587	447,417	Planned Completion Date is 2046
Mission Support	264,354	1,185,468	1,449,821	Planned Completion Date is 2046
HAMMER	72,682	173,122	245,804	Planned Completion Date is 2046
Solid Waste Storage and Disposal	336,193	732,934	1,069,127	Planned Completion Date is 2046
Analytical Services	315,871	830,764	1,146,634	Planned Completion Date is 2046
Total	**13,022,014**	**37,353,505**	**50,375,519**	

3.2.3 *Work Scope Summary*

The EM cleanup mission at the Hanford Site centers on the need to remedy the environmental contamination caused by decades of activities related to the production of plutonium. Having served as the nation's first full-sized plutonium production operation, Hanford's current projects are now specifically focused on minimizing, processing, and storing the backlog of radioactive and hazardous waste generated from 1943 through today; managing spent nuclear fuels and special nuclear material (SNM); decontaminating and decommissioning surplus facilities; and remediating the site.

The scope of work at the Hanford Site includes the management, cleanup, and disposition of soil, rubble, debris, and groundwater contaminated with radionuclides and hazardous substances as well as the management of high-level waste sludges, salts, and liquids. The sections below describe the major waste, material, and contaminated media volumes to be addressed by the Richland Operations Office. The volumes reported are approximate, and correspond to the major waste, material, and media flows, the potential treatment processes, and the off-site disposal destinations presented in Exhibit 3-10, the Richland Operations Office Conceptual Summary Disposition Map.

Transuranic Waste

- Approximately 16,000 cubic meters of legacy transuranic waste are currently in inventory and 6,500 cubic meters are expected to be generated over the life cycle of cleanup operations. After sorting and repackaging, approximately 17,000 cubic meters are planned to be disposed of at WIPP.

High-level Waste

- Approximately 220,000 cubic meters of high-level waste sludges, salts and liquids are currently contained in 149 single-shell and 28 double-shell holding tanks. After sludge washing, separation, and on-site vitrification, 14,000 cubic meters of waste are expected to be disposed of in an off-site geologic repository and 240,000 cubic meters are expected to be disposed of in an on-site low-level waste vault. Once empty, all holding tanks are expected to be stabilized and closed in place.

Other Waste

- Approximately 8,600 cubic meters of mixed low-level waste are currently in inventory and 64,000 cubic meters of mixed low-level waste are expected to be generated over the life cycle of cleanup operations. After treatment, 99,000 cubic meters of Hanford waste are expected to be disposed of on site.

- Approximately 180 cubic meters of low-level waste are currently in inventory and 130,000 cubic meters are expected to be generated over the life cycle of cleanup operations. An additional 32,000 cubic meters are expected to be received from DOE sites. After sorting, stabilization, and some commercial treatment, 230,000 cubic meters are expected to be disposed of on site.

Exhibit 3-10 Richland Operations Office Conceptual Summary Disposition Map

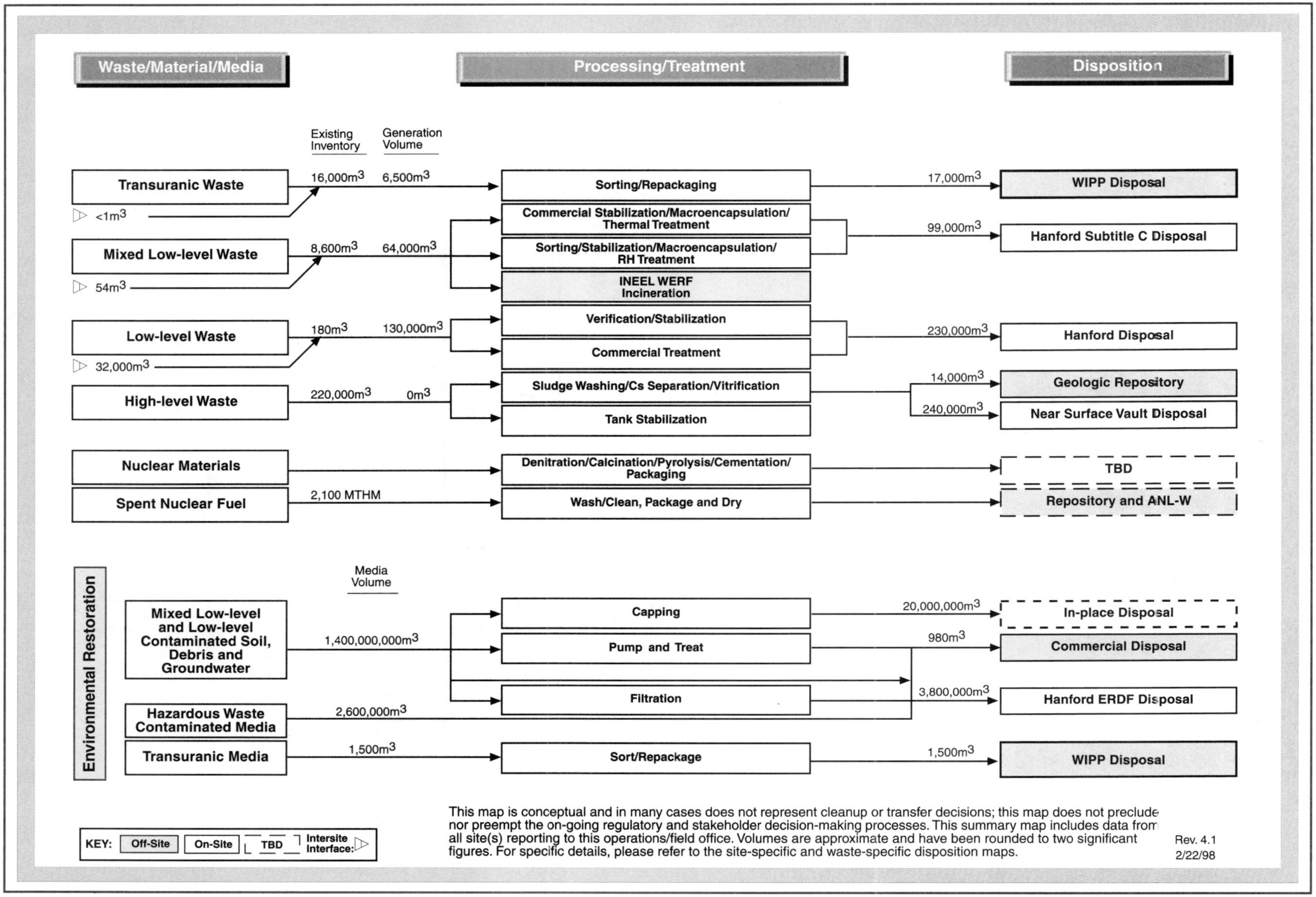

Remedial Action and Facility D&D

- Approximately 1.4 billion cubic meters of groundwater awaits a disposition decision, 20 million cubic meters of contaminated soil are expected to be capped in place, and 980 cubic meters of waste, consisting of spent resins generated from groundwater remediation and asbestos removed during deactivation and decommissioning of facilities, are expected to be disposed of at an off-site commercial disposal facility. Additionally, soils, rubble, and debris are expected to be disposed of at the ERDF.
- Approximately 1,500 cubic meters of debris contaminated with transuranic elements are expected to be generated during remediation activities. After sorting and repackaging, all 1,500 cubic meters are expected to be disposed of at WIPP.

Nuclear Materials

- Nuclear materials quantities are classified and cannot be disclosed in this document.

Spent Nuclear Fuel

- Over 2,100 metric tons heavy metal of spent nuclear fuel are currently in inventory. After washing, packaging, and drying, spent nuclear fuel is expected to be transferred to ANL-W or placed in a repository.

Exhibit 3-11 displays the Hanford Site closure costs by major work scope category. As depicted in the exhibit, the majority of the cost involved in the completion of environmental management activities at Richland revolves around high-level waste.

Exhibit 3-11
Richland Operations Office
Life-Cycle Costs by Category

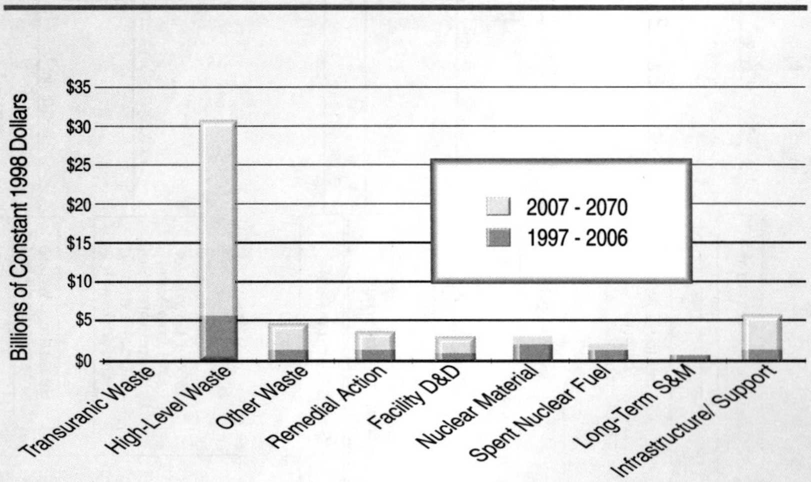

3.2.4 Critical Closure Path and Programmatic Risk

The critical closure path schedule presented in Exhibit 3-12 sets forth the timetable for completing closure activities at the Richland Operations Office. The Hanford Site critical closure path reflects those cleanup activities which are key to achieving completion of the site cleanup mission and end states. In Exhibit 3-12, the highlighted activities show the critical closure path, which represents the series of events that drive the overall completion date for the site; the bars represent projects and activities, and the diamonds represent critical events and milestones that must occur for Richland to be completed by 2046.

As shown in Exhibit 3-12, this path goes through the retrieval, treatment, and disposition of the high-level waste currently stored in the Hanford tanks. To succeed along this critical closure path, many other activities are also critical: (1) urgent risks must have top priority, (2) the fixed costs for maintaining the site in a safe manner need to be reduced through facility stabilization and deactivation to make additional funds available for cleanup, and (3) the Environmental Restoration Project must remain a high priority because it results in visible near-term cleanup progress. Another concern is that the practice of storing wastes awaiting treatment and deferring the retrieval and processing of the transuranic retrievable wastes eventually will increase costs for additional storage facilities.

Completion of the EM mission at the Richland Operations Office as scheduled will depend on the timely accomplishment of critical activities and events. Sites have assigned programmatic risk scores to each of the critical activities/milestones. Appendix D provides a complete definition of programmatic risk. Exhibit 3-12 illustrates that Hanford has twelve projects and their associated activities and milestones with high programmatic risk (programmatic risk scores of 4 or 5 in any category). Two of these twelve are on the critical closure path and are associated with the Tank Waste Remediation System project and the disposition of high-level wastes. As stated in the previous paragraph, there are a number of other activities that are not on the "critical closure path" but are necessary for success along the critical path. These activities include Spent Nuclear Fuel, Waste Management, Environmental Restoration, and Transition Projects. Each of these projects have high programmatic risks assigned to their associated activities and milestones. Exhibit 3-13 presents a summary of milestones and critical path activities with high programmatic risk.

This page intentionally left blank.

Exhibit 3-12

Richland Operations Office Critical Closure Path

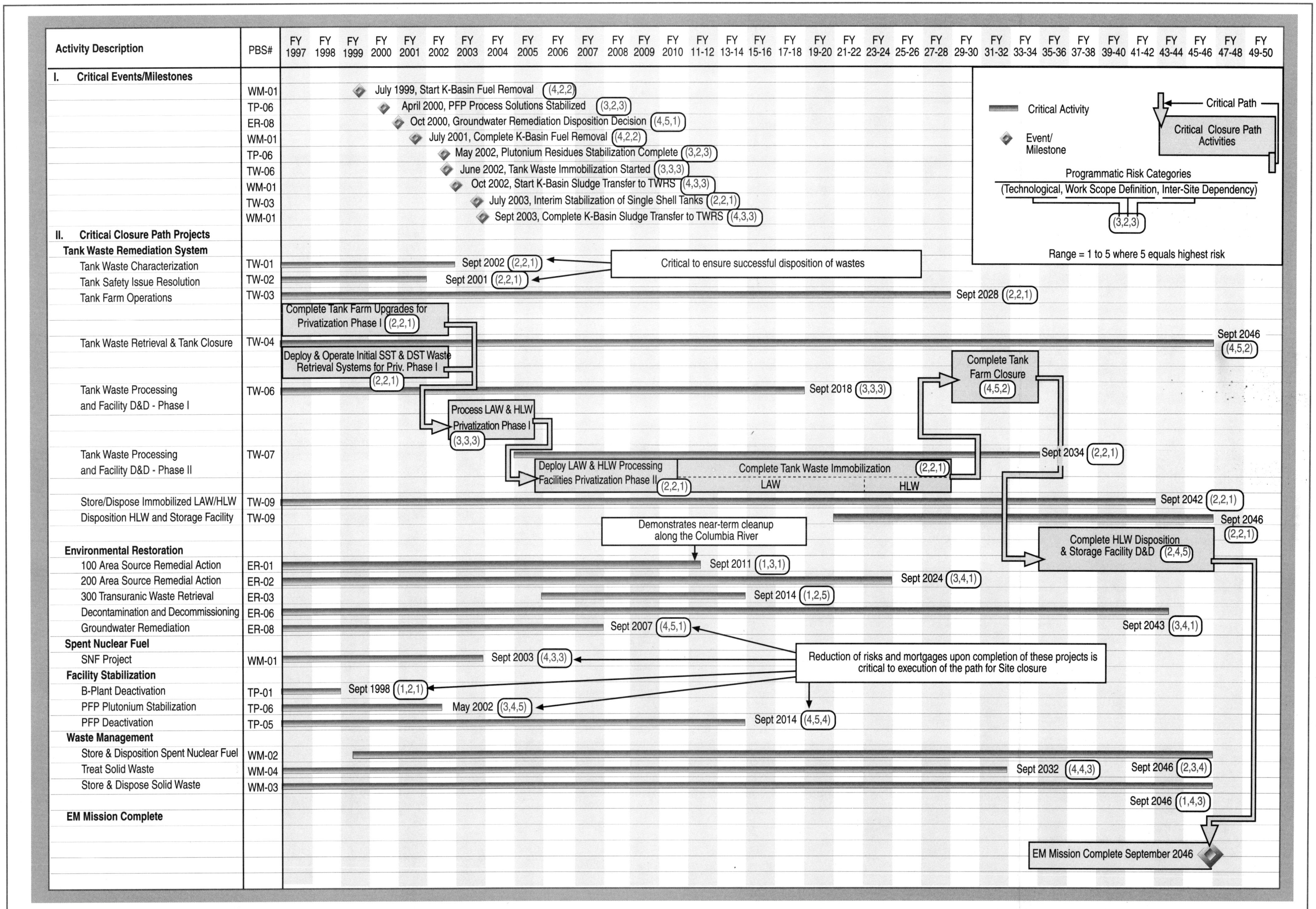

Exhibit 3-13

Summary of High Programmatic Risk Activities/Milestones: Richland Operations Office[a]

Project, Activity, Event	Start/ End Date	Programmatic Risk Categories		
		Technological	Work Scope Definition	Intersite Dependency
Start K-Basin fuel removal	Jul 99	4	2	2
Groundwater remediation disposition decision	Oct 00	4	5	1
Complete K-Basin fuel removal	Jul 01	4	2	2
Start K-Basin sludge transfer to TWRS	Oct 02	4	3	3
Complete K-Basin sludge transfer to TWRS	Sep 03	4	3	3
Complete Tank Farm closure	Sep 28/ Sep 34	4	5	2
Complete HLW disposition and storage facility D&D	Sep 34/ Sep 46	2	4	5
200 Area Source Remedial Action	Oct 97/ Sep 24	3	4	1
300 Area TRU Waste Retrieval	Sep 06/ Sep 14	1	2	5
Decontamination and Decommissioning	Oct 97/ Sep 43	3	4	1
PFP Plutonium Stabilization	Feb 97/ May 02	3	4	5
PFP Deactivation	Feb 97/ Sep 14	4	5	4

[a]Richland's critical closure path diagram (Exhibit 3-12) identifies additional high risk activities that are not on the critical path but are crucial for success along the critical closure path.

3.3 Savannah River Operations Office Summary

The Savannah River Site (SRS) was established in 1950 to produce special radioactive isotopes for national security purposes (e.g., plutonium-239 and tritium). In addition to this primary mission, SRS has produced other special isotopes to support research in nuclear medicine, space exploration, and commercial applications (for example, californium-252, plutonium-238, and americium-241).

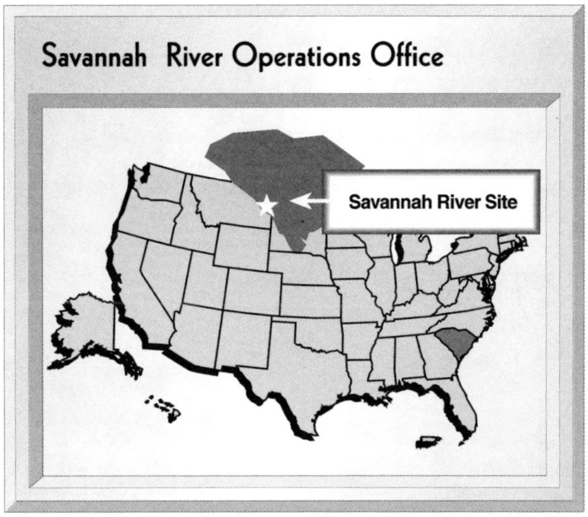

Since the end of the Cold War, the mission of SRS has changed. Emphasis has shifted from nuclear material production to environmental management. The Environmental Management (EM) program was initiated in 1989 to address the closure of old burial grounds and seepage basins. In FY 1992, the last of the production reactors was briefly operated. The production mission of the reactor program and supporting facilities was formally ended the following year. Current activities managed by EM cover three major programs: nuclear material and facility stabilization and facility deactivation; environmental restoration; and waste management. The primary drivers for these programs are the Federal Facility Agreement, the Federal Facility Compliance Act Consent Order, the Defense Nuclear Facilities Safety Board (DNFSB) Recommendation 94-1, and DOE order 430.1A (Life Cycle Asset Management). These agreements define commitments and milestones for the Savannah River Site.

3.3.1 End State

The status of the projects is such that no significant land use changes are projected through 2006. While progress will be made to reduce legacy risks and eliminate "mortgage" requirements as much as possible, the land-use designations will remain basically unchanged for any particular project area and the site as a whole. Significant changes in land-use designations may occur in the future, and will be addressed as the SRS Comprehensive Plan is developed. Development of this plan began in the fall of 1997, and will be completed in 10-14 months. Stakeholder involvement in future land-use decisions has already begun with the Savannah River Site Citizens Advisory Board, area planners, chambers of commerce, municipalities and others providing suggestions for future land use. As the Comprehensive Plan is developed, internal and external site stakeholders will be continually

involved in the process. SRS plans to store mixed waste off site at a Resource Conservation and Recovery Act (RCRA) Subtitle C Landfill once the mixed waste Record of Decision (ROD) is issued. SRS is planning to accept 473 spent nuclear fuel casks from foreign sources and 1,241 spent nuclear fuel casks from domestic sources during the entire spent fuel receipt program (1996 through 2035). The receiving basin for the fuel is expected to remain classified as nuclear industrial use.

After the site EM mission is complete, site boundaries should remain unchanged, and the land should remain under the ownership of the federal government for either a new site mission or as the first National Environmental Research Park. Regional environmental groups and national researchers have stressed that the site boundaries should remain unchanged to preserve its unique habitats. The flora and fauna at the site are such that the site could be used as a sanctuary for environmental study and observation. Additional information about Savannah River end states and long-term stewardship can be found in the Savannah River version of *Paths to Closure.*

3.3.2 Cost And Completion Dates

The Savannah River Operations Office has divided its environmental management work into 84 discrete projects. A Project Baseline Summary (PBS) exists for each project and contains detailed programmatic information, including cost, schedule, scope, end state, and interim milestones. A summary of the cost and schedule information for these projects is illustrated in Exhibit 3-14 (some of the 84 projects have been combined to simplify the graphic). For more information on each project, see the individual PBS.

The estimated EM life-cycle cost for the Savannah River Operations Office is $29.7 billion (constant 1998 dollars). This estimate does not include approximately $0.1 billion (constant 1998 dollars) of non-EM costs. The life-cycle cost is a planning estimate which includes costs for facility deactivation and long-term monitoring. Decisions on the ultimate end state of some of the facilities have not been made yet; the planning estimate is not intended to preclude any ultimate end state options. Based on these planning assumptions, the estimate could be applied to a range of decontamination and decommissioning options, including cocooning of facilities, as well as potential environmental restoration work. The overall completion date for EM work scope at the Savannah River Site is 2038, with long-term surveillance and monitoring activities continuing beyond 2070.

Exhibit 3-14 Savannah River Operations Office
Cleanup Project Summary: Duration and Costs (All costs in thousands of 1998 dollars)

Site Closure Project Activities	1997 - 2006	2007 - 2070	Total	
Savannah River Operations Office				
Savannah River Operations Office	163,980	261,900	425,880	Planned Completion Date is 2028
Savannah River Site (SRS)				
Ecology Lab Project	85,199	152,453	237,652	Planned Completion Date is 2030
DOE Projects Line Item	10,794	0	10,794	
Tank Farm Service Upgrades	13,395	0	13,395	
Canyon Exhaust Line Item	9,685	0	9,685	
H-Tank Farm Storm Water System Upgrades	11,563	0	11,563	
Tank Farm Support Services F-Area	31,037	0	31,037	
Actinide Packaging Line Item	180,765	0	180,765	
F-Area Stabilization Project	1,205,751	0	1,205,751	
H-Area Stabilization Project	1,289,751	0	1,289,751	
Deactivation Projects	0	347,598	347,598	
F-Tank Farm	545,752	656,577	1,202,330	
High-Level Waste System Upgrades	390,213	502,840	893,053	Planned Completion Date is 2023
H-Tank Farm	911,889	1,022,471	1,934,360	Planned Completion Date is 2024

Exhibit 3-14 Savannah River Operations Office (Continued)
Cleanup Project Summary: Duration and Costs (All costs in thousands of 1998 dollars)

Site Closure Project Activities	1997 - 2006	2007-2070	Total	97 98 99 00 01 02 03 04 05 06 07 08 09 10 11 12 13 14 15 16 17 18 19 20 21 22 23-70
In-Tank Precipitation (ITP)/ Extended Sludge Processing (ESP)/ Late Wash (LW) Operations	863,276	1,680,667	2,543,943	Planned Completion Date is 2024
Vitrification	1,425,110	2,556,850	3,981,961	Planned Completion Date is 2024
Saltstone	124,610	350,925	475,536	Planned Completion Date is 2024
Effluent Treatment Facility	200,997	371,042	572,039	Planned Completion Date is 2025
Waste Removal Operations and Tank Closure	111,221	684,077	795,298	Planned Completion Date is 2026
Glass Waste Storage	115,569	39,747	155,316	Planned Completion Date is 2026
Nuclear Materials Storage	120,455	439,780	560,235	Planned Completion Date is 2028
Savannah River Natural Resource Management & Research Institute	72,370	113,471	185,841	Planned Completion Date is 2030
Infrastructure Projects	432,466	689,576	1,122,043	Planned Completion Date is 2030
Solid Waste Projects	556,502	1,550,299	2,106,801	Planned Completion Date is 2030
Spent Nuclear Fuel Projects	920,958	1,151,339	2,072,297	Planned Completion Date is 2035
Environmental Restoration Projects	1,161,255	873,306	2,034,562	Planned Completion Date is 2038
Wackenhut Services - Incorporated SRS (WSI) Landlord Project	488,411	1,058,544	1,546,954	
F-Area Monitoring	267,136	1,467,136	1,734,272	
H-Area Monitoring and Minor Facility Monitoring	82,657	927,378	1,010,035	

Exhibit 3-14 Savannah River Operations Office (Continued)
Cleanup Project Summary: Duration and Costs (All costs in thousands of 1998 dollars)

Site Closure Project Activities	1997 - 2006	2007 - 2070	Total	97	98	99	00	01	02	03	04	05	06	07	08	09	10	11	12	13	14	15	16	17	18	19	20	21	22	23-70
M-Area Monitoring Project	76,485	115,378	191,863																											
D-Area Monitoring Project	500	22,687	23,187																											
Reactors Monitoring Project	93,704	619,337	713,042																											
Receiving Basin for Off-site Fuels Monitoring Project	0	76,017	76,017																											
Depleted Uranium Storage																	Project Currently Not Funded													
Total	**11,963,454**	**17,731,398**	**29,694,852**																											

The projected cost profile for EM activities associated with the Savannah River Operations Office was developed by combining the cost estimates presented in each of the Project Baseline Summaries. Exhibit 3-15 displays the resultant baseline cost profile.

Exhibit 3-15
Savannah River Operations Office
Environmental Management Baseline Cost Profile

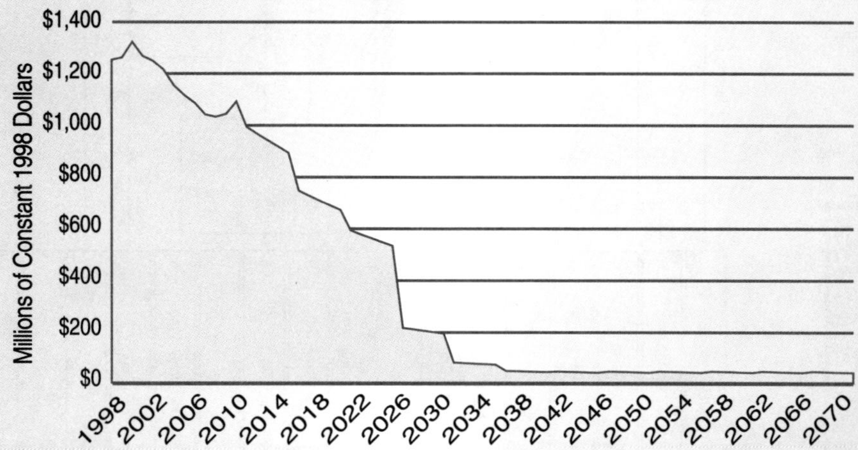

3.3.3 *Work Scope Summary*

The scope of work at the Savannah River Operations Office includes the management of high-level waste sludges and salts; spent nuclear fuel from DOE facilities, universities, and foreign research reactors; soil, sludges, debris, and groundwater contaminated with radionuclides and hazardous substances; and numerous excess facilities and nuclear materials. The sections below describe the major waste, material, and contaminated media volumes to be addressed by the Savannah River Operations Office. The volumes represented are approximate, and correspond to the major waste, material, and media flows, the potential treatment processes, and the off-site disposal destinations presented in Exhibit 3-16, the Savannah River Operations Office Conceptual Summary Disposition Map.

All waste and material existing at the Savannah River Site, as well as the waste generated from the cleanup process itself, will be managed as described in the Savannah River Operations Office version of *Paths to Closure* and the Project Baseline Summaries.

Exhibit 3-16 Savannah River Operations Office Conceptual Summary Disposition Map

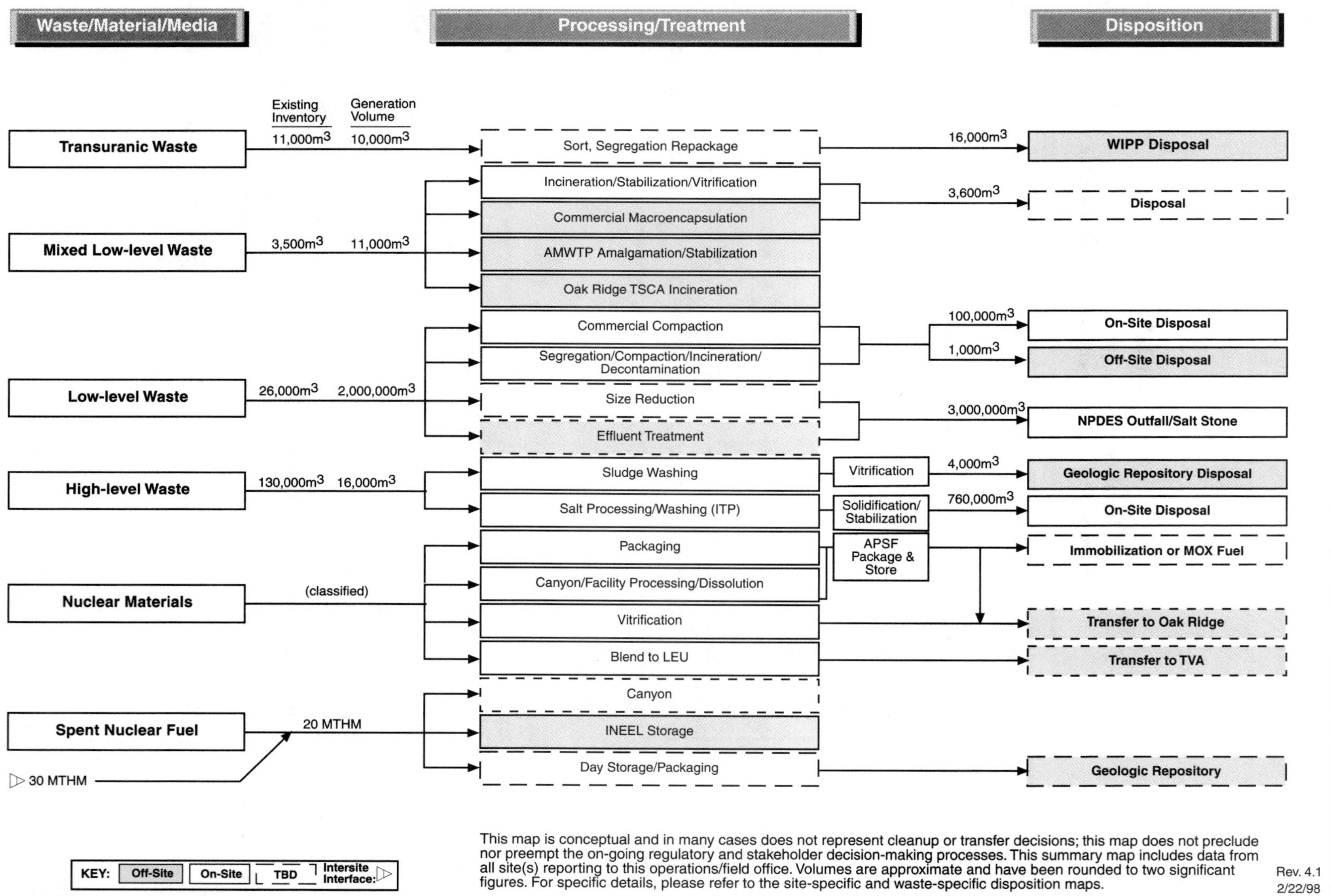

This map is conceptual and in many cases does not represent cleanup or transfer decisions; this map does not preclude nor preempt the on-going regulatory and stakeholder decision-making processes. This summary map includes data from all site(s) reporting to this operations/field office. Volumes are approximate and have been rounded to two significant figures. For specific details, please refer to the site-specific and waste-specific disposition maps.

Rev. 4.1
2/22/98

Exhibit 3-16 Savannah River Operations Office Conceptual Summary Disposition Map (Continued)

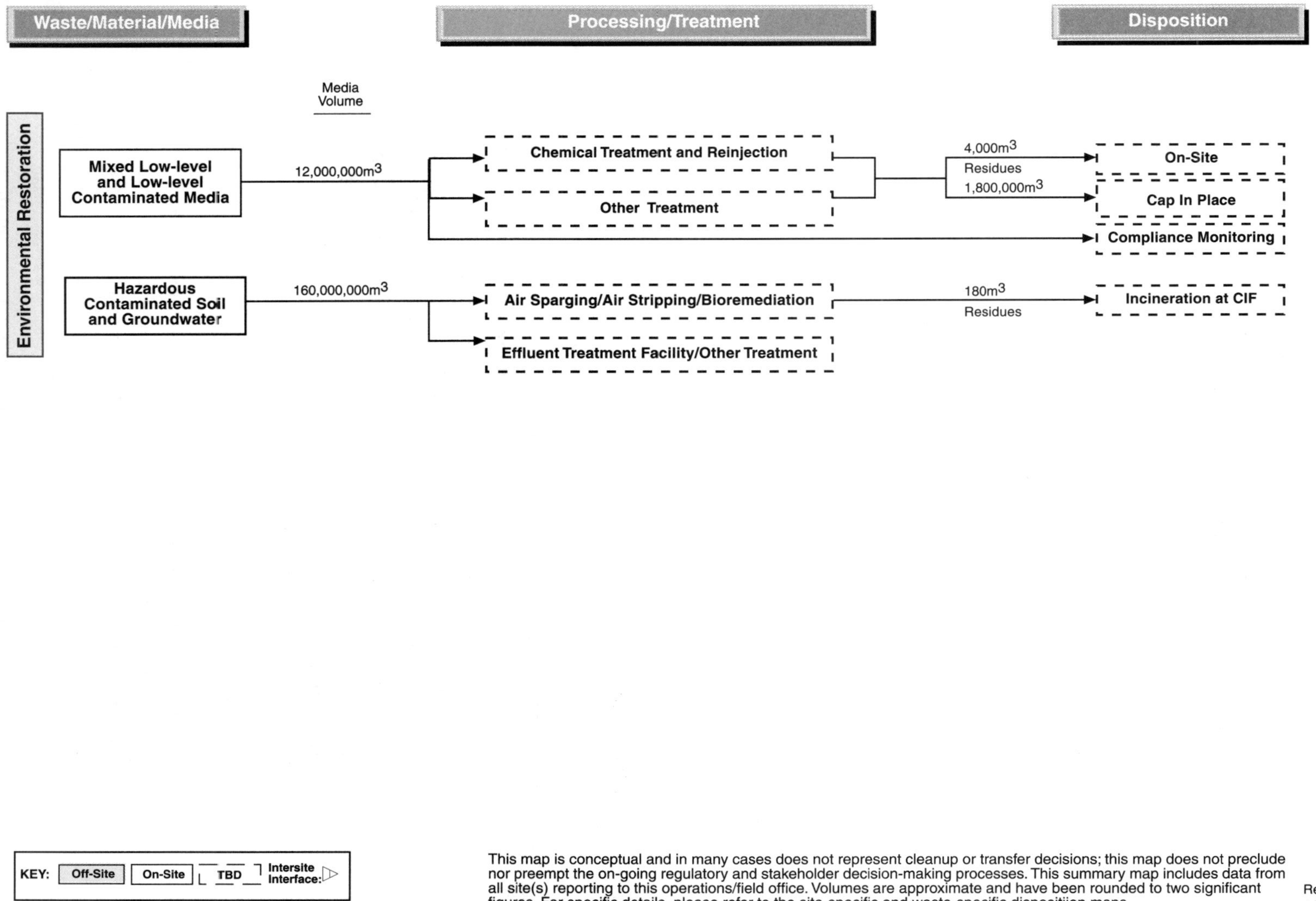

This map is conceptual and in many cases does not represent cleanup or transfer decisions; this map does not preclude nor preempt the on-going regulatory and stakeholder decision-making processes. This summary map includes data from all site(s) reporting to this operations/field office. Volumes are approximate and have been rounded to two significant figures. For specific details, please refer to the site-specific and waste-specific dispositiion maps.

Rev. 4.0
2/20/98

Transuranic Waste

- Approximately 11,000 cubic meters of transuranic waste are currently in inventory (primarily stored in drums and black boxes) and 10,000 cubic meters of transuranic waste are expected to be generated over the life cycle of cleanup operations. After a combination of sorting, segregation, and repackaging, 16,000 cubic meters are planned for disposal at WIPP.

High-level Waste

- Over 130,000 cubic meters of high-level waste are currently in inventory and approximately 16,000 cubic meters of high-level waste are expected to be generated from future nuclear material separation operations. After sludge washing, salt processing, and vitrification, 4,000 cubic meters of vitrified high-level waste are planned to be disposed of at an off-site geologic repository and 760,000 cubic meters of low-level waste saltstone are planned to be disposed of at an on-site vault.
- Forty-nine high-level waste tanks and additional facilities will be managed. After washing and stabilization, tanks will be closed in place and other facilities will be deactivated.

Other Waste

- Approximately 3,500 cubic meters of mixed low-level waste are currently in inventory and over 11,000 cubic meters of mixed low-level waste are expected to be generated over the life cycle of cleanup operations. After a range of treatment activities, 3,600 cubic meters are expected to be disposed of at an off-site facility.
- Approximately 26,000 cubic meters of low-level waste are currently in inventory and over 2.0 million cubic meters of low-level waste (including 1.3 million cubic meters of process water) are expected to be generated over the life cycle of cleanup operations. After a range of treatment activities, including effluent treatment and commercial compaction, 100,000 cubic meters are expected to be disposed of at an on-site disposal cell, 1,000 cubic meters are expected to be sent to an off-site commercial facility, and 3.0 million cubic meters of treated effluent are planned to be discharged through a National Pollutant Discharge Elimination System outfall.

Remedial Action

- Approximately 12 million cubic meters of environmental media including soil, rubble & debris, and groundwater contaminated with radionuclides and hazardous substances will be managed. After treatment, 4,000 cubic meters of residues are expected to be disposed of on site and 1.8 million cubic meters of environmental media are expected to be capped in place.
- Nearly 160 million cubic meters of environmental media, including soil, rubble and debris, and groundwater contaminated with hazardous substances, will

be managed. In addition to the planned incineration of 180 cubic meters of residues at the Consolidated Incineration Facility (CIF), contaminated media are expected to be addressed by a number of treatment processes, including air sparging and air stripping.

Nuclear Materials

- Nuclear materials quantities are classified and cannot be disclosed in this document.

Spent Nuclear Fuel

- Approximately 20 metric tons heavy metal of spent nuclear fuel are in inventory and 30 metric tons heavy metal of spent fuel are expected to be received from off site. After on-site management, the spent fuel is expected to be placed in an off-site geologic repository.

Exhibit 3-17 illustrates the life-cycle costs by major work scope categories. High-level waste accounts for the largest portion of the total life-cycle cost at the Savannah River Operations Office. The Facility Deactivation category accounts for the second greatest portion of life-cycle costs.

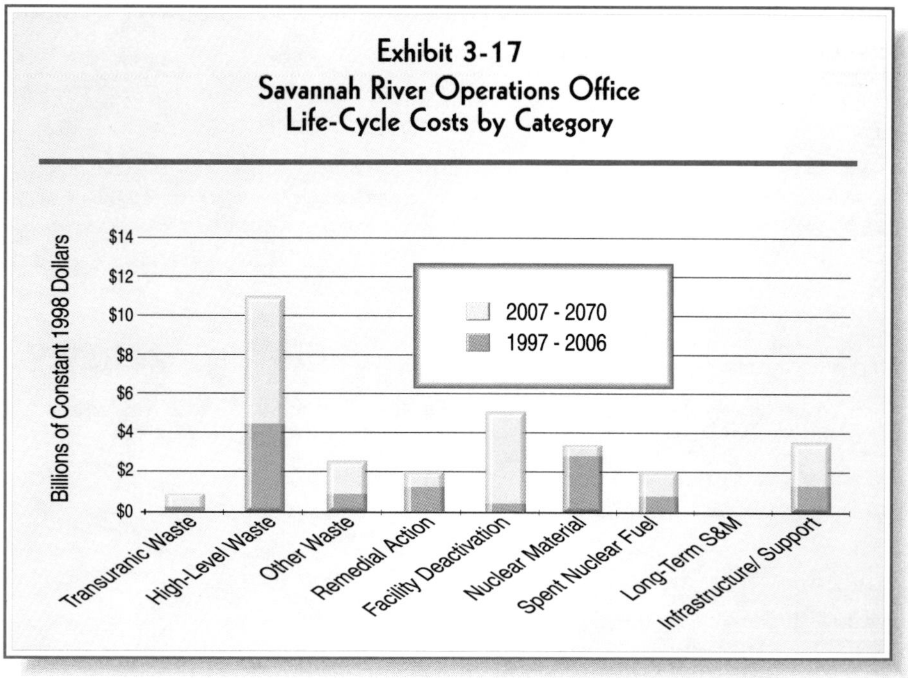

3.3.4 Critical Closure Path and Programmatic Risk

The critical closure path schedule presented in Exhibit 3-18 sets forth the timetable for completing closure activities at the Savannah River Operations Office. The critical closure path identifies the sequence of major cleanup activities that have little scheduling flexibility and must occur without delay if the SRS EM cleanup mission is to be completed on time. In Exhibit 3-18, the highlighted activities show the critical closure path, which represents the series of events that drive the overall completion date for the site; the bars represent critical activities; and the diamonds represent critical events and milestones that must occur for Savannah River Operations Office to be completed as planned. Sites have assigned programmatic risk scores to each of these activities and events.

Completion of the EM mission at the Savannah River Operations Office as scheduled will depend on the timely accomplishment of critical activities and events. Exhibit 3-19 presents a summary of activities and milestones on the critical closure path that have high programmatic risk (programmatic risk scores of 4 or 5 in any category). Appendix D provides a complete definition of programmatic risk. In their formal PBS submission, Savannah River identified 22 activities and events with high programmatic risk values. Four of these have high work scope uncertainty and are associated with projects that have life-cycle costs in excess of one billion dollars. For more information on the management approach for these programmatic risk issues, see the Savannah River Operations Office version of *Paths to Closure.*

Exhibit 3-18 **Savannah River Operations Office Critical Closure Path** (Critical Activities and Critical Path)

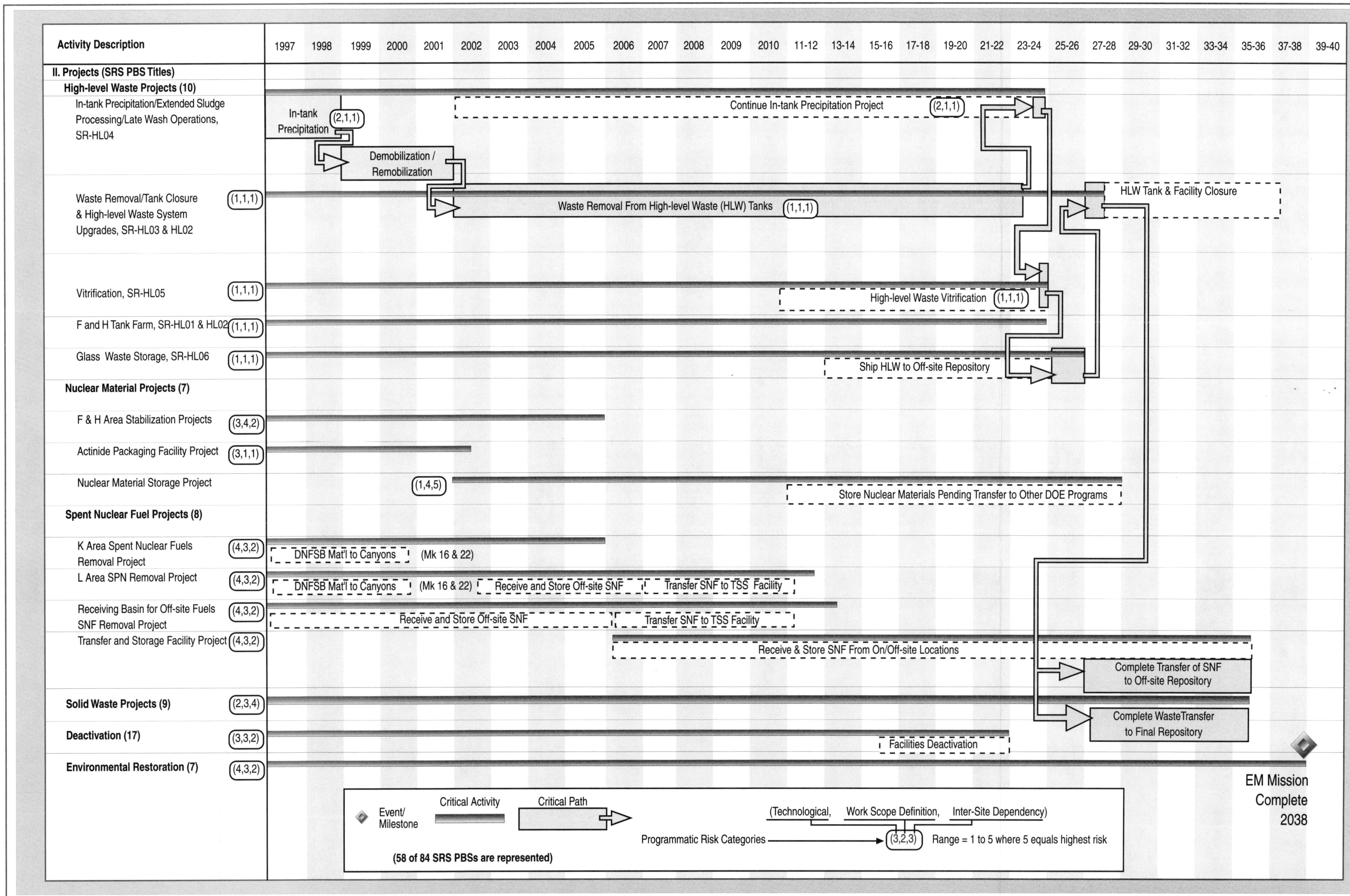

Exhibit 3-18 **Savannah River Site Critical Closure Path (Continued)** (Key Milestones)

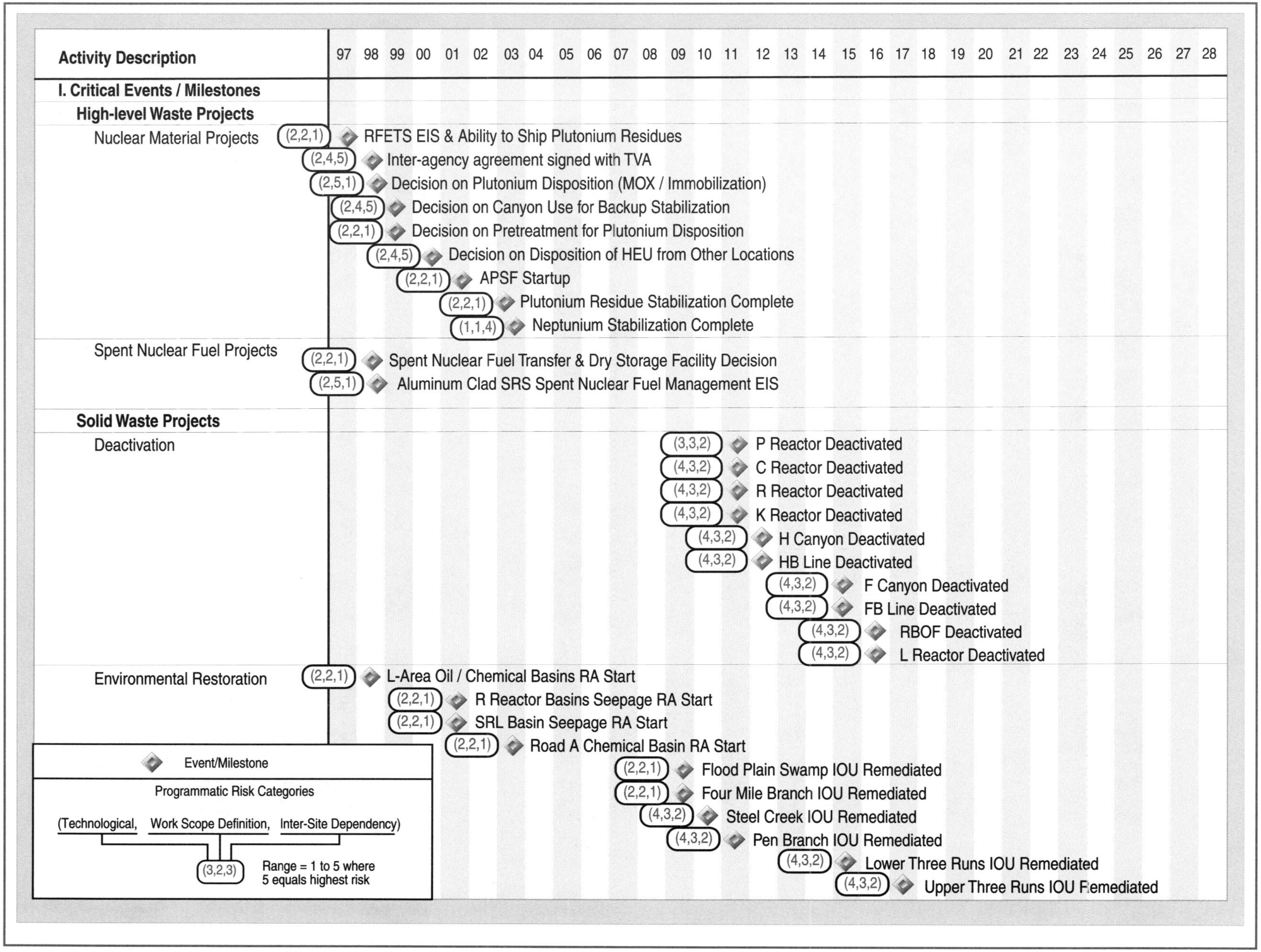

Exhibit 3-19
Summary of High Programmatic Risk Activities/Milestones: Savannah River Operations Office[a]

Project, Activity, Event	Start/ Completion Date	Programmatic Risk Categories		
		Technological	Work Scope Definition	Intersite Dependency
Inter-agency agreement signed with TVA	Jan 97/ Sep 98	2	4	5
Decision and disposition of HEU from other locations	Jan 97/ Sep 00	2	4	5
Np stabilization complete	Jan 97/ Nov 03	1	1	4
Four Mile Branch IOU Remediated	Oct 97/ Mar 09	4	3	2
Flood Plain Swamp IOU Remediated	Jun 00/ Apr 09	4	3	2
Steel Creek IOU Remediated	Nov 98/ Dec 10	4	3	2
R Reactor Deactivated	Jan 06/ Dec 11	4	3	2
K Reactor Deactivated	Jan 06/ Dec 11	4	3	2
P Reactor Deactivated	Jan 06/ Dec 11	4	3	2
C Reactor Deactivated	Jan 06/ Dec 11	4	3	2
HB Line Deactivated	Jan 06/ Dec 12	4	3	2
H Canyon Deactivated	Jan 06/ Dec 12	4	3	2
Pen Branch IOU Remediated	Dec 99/ Dec 14	4	3	2
Lower-Three Runs IOU Remediated	Oct 97/ Jun 15	4	3	2
F Canyon Deactivated	Jan 06/ Dec 15	4	3	2
FB Line Deactivated	Jan 06/ Dec 15	4	3	2

Summary of High Programmatic Risk Activities/Milestones:
Savannah River Operations Office (Continued)

Project, Activity, Event	Start/ Completion Date	Programmatic Risk Categories		
		Technological	Work Scope Definition	Intersite Dependency
RBOF Deactivated	Jan 06/ Dec 16	4	3	2
L Reactor Deactivated	Jan 06/ Dec 16	4	3	2
Upper Three Runs IOU Remediated	Jan 97/ Sep 17	4	3	2
Finish shipping vitrified waste to Federal Repository	Oct 24/ Sep 25	1	4	5
Interim SNF dry storage, conditioning, treatment, packaging and shipping facility	Sep 05/ Sep 35	3	4	4
Complete surveillance and maintenance of remediated waste units	Oct 96/ Sep 38	4	3	2

[a]Savannah River's critical closure path (Exhibit 3-18) identifies 13 high risk activities / milestones that were not identified in their formal PBS submission.

Chapter 4

Meeting Programmatic Challenges

Chapters 2 and 3 outlined a massive environmental management cleanup program, the cost of which is an estimated $147 billion (constant 1998 dollars). Completion of the scope of work of the program will take more than 50 years. To reduce the monumental costs of the cleanup effort, Environmental Management (EM) sites must seek, and find, significant opportunities to accelerate the scope of work of the cleanup. *Paths to Closure*, while grounded in baseline estimates, explores opportunities to increase efficiency and thereby enhance performance that will enable the EM program to achieve its cleanup mission more quickly and at a lower cost.

EM's adoption of such opportunities to enhance performance is the first step in resolving problems that will arise because of inevitable differences between baselines and either assumed or actual funding levels for any given year. *Paths to Closure* also outlines other options for reducing life-cycle costs, should enhanced performance not address fully the funding challenges that an effort of the size of EM's cleanup program will face.

Since EM began developing the vision of accelerated cleanup, the President and Congress have reached a balanced budget agreement. As an underlying premise, therefore, *Paths to Closure* reflects the Department of Energy's (DOE's) need to control costs and comply with the President's balanced budget agreement with Congress. Consistent with this premise, DOE's annual budgeting process includes a process for making adjustments to account for differences between work that is planned, annual appropriations, and projected funding levels using information contained in *Paths to Closure*.

4.1 Relationship Between Baselines and Funding Guidelines

In developing the estimates of cost and schedule set forth in Chapters 2 and 3, the EM program assigned each Operations/Field Office an annual funding guideline which was consistent with recent appropriations levels. In some cases, sites exceeded this funding guideline to meet compliance commitments. EM established the funding guideline last October prior to receiving final FY 1999 and outyear budget targets. It was essential to establish an assumption at that time in order to produce a draft of this report by February 1998. For planning purposes, this funding assumption has not changed.

The EM program assumed that the $5.75 billion per year funding guideline would already include adjustments for inflation—the same assumption the federal government makes in providing outyear budget targets to government agencies for planning. In effect, the true buying power of the EM program decreases over time. In developing their baselines, each Operations/Field Office factored the effect of inflation into planning assumptions as they scheduled work.

The funding guideline can be compared with the baseline in one of two ways: current year dollars (that is, dollars that include costs associated with inflation), or constant 1998 dollars (that is, dollars that have been adjusted to remove the inflationary component, in the manner in which data are reported in Chapters 2 and 3 of this document). Exhibit 4-1 illustrates the correct comparison of the funding guideline with the baseline using both current and constant 1998 dollars. As the exhibit shows, EM's overall baseline, which has not been adjusted to reflect FY 1998 appropriations and the FY 1999 budget request, exceeds the funding guideline from the current period through 2006. The projected difference during the period 1999 to 2006 is estimated at $4.4 billion in current year dollars or $3.9 billion in constant 1998 dollars. At this time, the forecasted difference over the next eight years is only an estimate, but highlights the need to maximize enhanced performance and work with stakeholders, regulators, and Tribal Nations to review site priorities and identify the best use of resources under various funding scenarios.

Exhibit 4-1
Comparison of the Baseline to the Assumed Funding Level

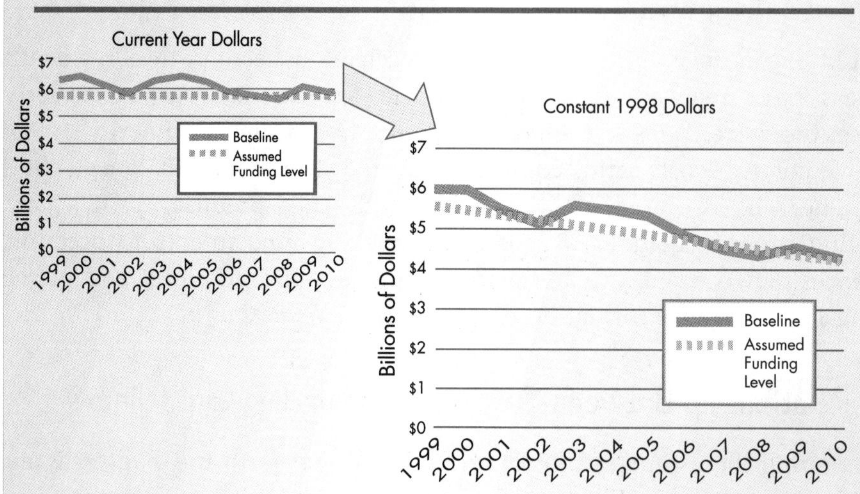

To facilitate a better understanding of what drives the baseline requirements and funding needs for the near term, EM has identified requirements drivers. EM uses the requirements drivers during the annual budget process to identify program needs in detail. The overall baseline cost for EM is driven largely by four components:

(1) **Compliance.** Compliance activities are those designed to meet all legally applicable requirements as directed by Executive Order 12088. During the annual budget process, EM asks sites to identify funding requirements to meet compliance agreements, court orders, settlement agreements, consent decrees, federal, state, and local statutes or regulations. Compliance by far accounts for the largest cost element of the program. For FY 1999, baseline estimates include $5.1 billion for compliance costs.

(2) **Additional "Minimum-Safe" Activities.** Site baseline estimates also reflect the scope, schedule, and costs necessary to conduct "minimum-safe" activities, which are necessary to address recommendations of the Defense Nuclear Facilities Safety Board (DNFSB) and comply with applicable DOE Orders that ensure the safety and health of workers and the public, and protection of the environment. Many "minimum-safe" activities actually are included in compliance activities. Baseline estimates include approximately $122 million for requirements that result strictly from DNFSB commitments and compliance with DOE Orders for safety and health in addition to the $5.1 billion earmarked for compliance.

(3) **Additional High-Priority Items.** Site baseline estimates may include additional high-priority work scope including program management and support activities, planning and oversight functions, and other activities associated with the management and completion of work under the EM program. For FY 1999, such high-priority items are estimated to account for approximately $156 million of the overall baseline total, in addition to the $5.1 billion for compliance and $122 million for additional "minimum-safe" activities. The costs of accelerated closure activities at the Rocky Flats Environmental Technology Site and the Ohio Field Office sites also are included in this category.

(4) **National Programs, Federal Salaries, and Headquarters Functions.** A portion of the overall baseline estimate of the cost of the EM program includes National Program activities, salaries for federal employees who oversee work in the Field, and other crosscutting work that supports the effective execution of EM's responsibilities. Specifically, such activities include the National Science and Technology program, the National Transportation program, and the National Pollution Prevention program. Most such activities are executed in the Field; EM Headquarters provides oversight, overall management, and policy guidance. For FY 1999, the estimate of baseline costs to support the activities of Headquarters and the National Programs is approximately $627 million.

Exhibit 4-2 displays a breakdown of the baseline cost of the EM program by the four categories discussed above over time. Because such data are collected only for the budget planning year, the exhibit is based on the assumption that the trend for FY 1999 will continue through time.[6] As the graph shows, for several

[6]It is very difficult to estimate compliance requirements in detail for outyears. Many compliance agreements have two- to three- year windows within which requirements are specified; definitive needs beyond that window have not been fully documented. At other sites, compliance requirements are defined more fully. For analysis at the EM level, Exhibit 4-2 simply extrapolates compliance needs based on FY 1999 data. This methodology provides a high-level mechanism for comparing compliance needs with potential planning levels.

years between the current period and FY 2006, there is the potential that the EM program will experience a difference between the funding guideline of \$5.75 billion per year and the baseline estimate.

A closer examination of Exhibit 4-2 shows that, even if the focus were on compliance alone, the difference would remain for some years (assuming that National Programs and federal oversight activities are funded).

Exhibit 4-2
Comparison of Assumed Funding Level to EM Baseline (constant 1998 dollars)

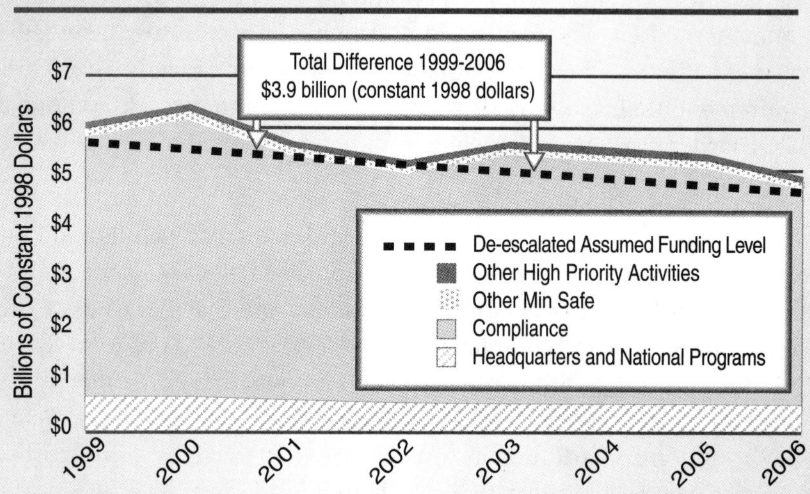

4.2 Reducing Costs and Maintaining Schedules

Paths to Closure is not a budget or decision document. The annual budget process is different from *Paths to Closure.* Establishing the EM budget requires a careful balancing of multiple factors:

- Protecting public health, the environment, and workers;
- Complying with applicable environmental laws, regulations, and agreements;
- Accelerating the completion of cleanup activities at DOE sites;
- Allocating resources among DOE sites;
- Weighing EM program needs against competing DOE and Executive Branch needs such as the President's recent balanced budget agreement with Congress; and
- Accounting for "local" priorities of stakeholders and Tribal Nations at individual sites.

Although *Paths to Closure* is not part of the annual budget development process, the two are related. *Paths to Closure* is a useful tool, not only for assisting in annual budget formulation, but also for making annual adjustments to the execution of the cleanup program based on budget funding decisions. In evaluating annual budget scenarios, *Paths to Closure* gives EM the management tools needed to understand impacts to life-cycle costs and closure date schedules.

Paths to Closure is representative of baselines and is not updated to reflect various budget scenarios that occur throughout the course of the annual budget process. This is because it is extremely difficult and unrealistic for sites to "re-baseline" multiple times during the course of a year. Typically, EM works to align the baselines on a year-to-year basis so that work scope planned for the execution year (currently FY 1998) is consistent with the budget. During these annual updates, sites can also reflect outyear planning changes in the baseline, changes that have resulted from variances in actual results from the previous year, scope changes, enhanced performances, improvements in estimates, and other changes in planning assumptions.

The Environmental Management program recognizes that there will be differences in future iterations of *Paths to Closure* between actual budget requests and appropriations and the funding guideline amount due to the dynamic nature of the budget process. Because of the inevitability of differences between baselines, planning levels, and budget funding, the budget process contains a systematic process for resolving funding differences. Reducing life-cycle costs through enhanced performance, and therefore addressing differences between planning and funding levels, is EM's most viable and most desirable option. Receiving sufficient funds to eliminate all future differences is unlikely, given that DOE's costs must be controlled to meet the President's balanced budget agreement with Congress. The budget process includes a systematic process for making work execution adjustments to account for annual fluctuations in funding levels using information in *Paths to Closure*:

- Constantly seeking ways to enhance performance;
- Requesting additional funding and/or considering reallocation of funds among sites to address immediate health and safety needs;
- In cases of small funding differences in budget outyears, using funding available for other EM programs at a site to address compliance-related project scope; and
- In cases where large funding differences are projected, working with the Office of Management and Budget and the Congress to seek additional funds, and also working with stakeholders, regulators, and Tribal Nations to review sites' environmental activities to reach agreement on site programs that balance many competing priorities and needs.

The following sections discuss the steps of this part of the budget process in greater detail.

4.2.1 *Enhanced Performance Mechanisms*

Enhancing performance is not a strategy reserved for situations in which there is a funding issue; rather, it is an integral part of the overall EM program's work culture. EM has and will continue to implement performance enhancements as a means of reducing the significant costs of the cleanup program. The EM program has available a number of mechanisms or tools that offer the potential to reduce the life-cycle cost of the cleanup program and thus help address funding differences. These tools include the application of science and technology deployment, project sequencing, pollution prevention, contract reform, integration, and implementing lessons learned. As sites identify and document project-specific applications of these tools, the baselines will be modified to reflect the "real" savings, and permit the acceleration of other projects.

Application of Science and Technology Deployment. As the cleanup program has progressed, EM has accelerated the use of new technologies. Technology offers the potential to provide solutions to currently intractable problems and may offer better, safer, and cheaper alternatives to current baseline technologies. New technologies range in size from small thumbnail-size sensors that fit in one-inch pipes to the melter placed in the Defense Waste Processing Facility. These new technologies already are having a positive effect on the progress of cleanup. By 1998, more than 140 new technologies had been used to characterize and treat waste and to remediate contaminated soils and groundwater. As it is proven that such new technologies can lower cost while improving worker safety and reducing environmental risk, their use will increase.

Site versions of *Paths to Closure* have identified 543 science and technology needs based upon the designation of technical programmatic risk in the projects. The EM program intends to bring more than 100 new technologies to bear in the next four years to begin to address these needs. Each of the Operations/Field Offices has developed a site-specific technology deployment plan which describes its approach to overcoming barriers to technology deployment. Implementation of these plans will enable rapid integration of these new technologies into site cleanup activities to fill key technology gaps. The Accelerated Site Technology Deployment program, authorized by Congress for the first time in FY 1998, is a positive step towards that goal. This program accounted for 14 of the deployment opportunities identified in *Paths to Closure.*

EM has identified technology-related cost savings opportunities exceeding $9 billion. Of this amount, about $5 billion already has been incorporated into the assumptions used to develop site baseline estimates. However, some of the assumptions about technologies that have been incorporated into baselines require additional investment of resources to ensure their deployment. The additional benefits of innovative technologies presumably will be reflected in future baseline estimates as sites identify opportunities to use those technologies.

The budget requests for the Technology Development program for FY 1999 and FY 2000 were formulated and prioritized using the Operations/Field Office data provided through *Paths of Closure.* Each of the technology work packages is linked to, and prioritized by, specific EM projects and waste streams.

To reduce the cost of cleanup—and in some cases to allow cleanup—EM must identify, develop, and apply science and new technologies aggressively. In 37 Project Baseline Summaries (PBSs), representing an estimated life-cycle cost of $33 billion, more than 80 opportunities have been identified to help meet EM's enhanced performance goals. The potential savings identified for those 37 PBSs exceed $4 billion. Clearly, even a small fraction of the 353 projects discussed in *Paths to Closure* can contribute significantly to the achievement of enhanced performance goals. For the most part, savings associated with technology-based opportunities are to be realized in the high-level waste programs at the Savannah River and Richland Operations Offices.

Focusing Science and Technology Investments

The cleanup strategy aids EM in efforts to maximize return on investments made in science and technology.

For each project, sites have determined specific technology needs that could improve cleanup, accelerate the schedule, or reduce costs. Information about where and when new technologies are being deployed, a nationally prioritized set of technology needs, specific cost savings opportunities, and an assessment of technological risk are all crucial to building the right cleanup investment portfolio. EM will develop its science and technology budget based on these data. Such highly focused investments will help achieve challenging enhanced performance goals and reduce the technological risk associated with projects that are on the critical path to site completion.

The construction of science and technology roadmaps within EM and elsewhere in the Department will enable EM to bring the relevant research and development efforts of the rest of the Department to bear on EM's long-term, high cost projects, as well as high-risk activities and waste streams. The overall EM investment strategy for science and technology will be described in the *EM Research and Development Program Plan* which is scheduled to be released later this year.

EM has identified 50 PBSs that present medium to high technological risk that are on the critical path to site closure. The projects include more than 80 medium to high-risk activities or events that could benefit from highly focused investments in science and technology. EM will evaluate these high-risk projects carefully and identify those cases in which failure to complete the project will have the most significant effect on the progress of the cleanup program. EM-built, project-level roadmaps will be considered for those selected projects that can benefit from significant investments in science and technology.

In addition, through preparation of the disposition maps, the EM program has identified more than 80 waste streams that present medium to high technological risk. Disposition maps will also help to focus future science and technology investments. "Roadmapping" the technology needs and technological risk to specific science and technology investments will ensure that waste treatment can proceed successfully on the national level, according to an established process. The roadmaps will help establish requirements, both schedule and technical, for when and where the results of these investments need to be delivered.

Opportunities for technology-based cost savings identified in *Paths to Closure* represent an appropriate first step. However, as part of EM's roadmapping efforts, we will reevaluate the technical approach on long-term, high-cost activities that present minimal technological risk. More than 60 projects will extend past 2004, cost more than $50 million each, and present minimal technological risk. The EM program will review these projects to determine whether new technologies can replace conventional cleanup technologies to reduce costs and accelerate cleanup schedules.

Integration. Although each DOE site and laboratory is unique in its capabilities, some problems are common throughout the complex: e.g., what is the best technology to treat, store, and dispose of various types of radioactive and hazardous waste and how should we manage our nuclear materials inventory? Accordingly, EM will be utilizing existing unique capabilities and developing new technologies at sites to do business efficiently to achieve common objectives.

This integration effort means sharing across sites: consolidating treatment, storage, and disposal facilities where it makes good sense; applying innovative technologies among sites; and working to ensure consistency in reporting data such as waste inventory and generation, as well as available packaging and transportation systems for shipments of waste and nuclear materials.

The guiding planning document for DOE is the Strategic Plan. The Environmental Management program plays a key role in implementing the strategies and achieving the goals in the Strategic Plan. *Paths to Closure* provides more detail on the strategies being employed to meet the Department's strategic objectives. As strategies are developed, the EM program identifies gaps and opportunities for improvements. Integration provides valuable insight into ways to improve current strategies as well as proposed solutions which use resources effectively.

One of the first steps in the analysis of opportunities for integration is the uniform reporting of waste volumes and related data. Waste and material disposition maps are a new management tool added in response to stakeholder and Tribal Nation concerns about nuclear material and waste disposition. The maps are graphical representations of each DOE site's current conceptual approach to managing wastes, nuclear materials, and contaminated media from its current status through its ultimate disposition, including shipping and off-site treatment and disposal. Chapter 3 and Appendix E display Conceptual Summary Disposition Maps for each Operations/Field Office.

Project Sequencing. Projects for which "mortgages" or carrying costs are high typically include "support" activities, such as general maintenance, security, infrastructure, and other activities not directly associated with environmental cleanup. The scope and life-cycle cost of such projects could be reduced if the EM program were to accelerate their completion. EM has identified two general approaches to accomplishing "mortgage reduction": (1) increasing near-term investment in specific projects to allow for accelerated completion of those projects, and (2) reallocating funding to focus funds used at sites on projects with high "mortgages".

Mortgage and Mortgage Reduction

"**Mortgage**" refers to support activities and their associated costs. Mortgage costs represent the fixed-cost portion of a project and support activities required to maintain a facility and stored waste or material in a stable or operative configuration.

"**Mortgage Reduction**" refers to those activities whose primary focus is to treat waste, stabilize nuclear materials, and deactivate, decontaminate, and decommission facilities, and their associated costs. As such activities are completed, their associated mortgage costs are reduced or eliminated.

The EM "mortgage reduction" initiative has four objectives: (1) identify projects for which support costs are high (such as materials for stabilization, waste treatment or disposal, facility deactivation) and where acceleration of activities may reduce costs for support activities significantly; (2) identify those projects that offer a high potential internal rate of return if funding can be increased and if the "mortgage reduction" could be quantified; (3) identify those projects that currently are providing "mortgage reduction" benefits and quantify those benefits; and (4) identify those long-term, high cost projects that present minimal technological risk so that new technology can be applied to accelerate cleanup or reduce costs with minimal additional programmatic risk. In many cases, sequencing projects that have a high "mortgage reduction" potential also reduces urgent risks and meets our compliance commitments. By reducing high "mortgages", the EM program can reduce risk, accelerate site closures, minimize the need for near- and long-term surveillance and monitoring activities, and reduce support costs.

Pollution Prevention. The DOE pollution prevention program is a management tool for optimizing waste reduction and pollution prevention. Pollution prevention is a core program that helps sites maximize their environmental compliance, while reducing costs associated with the generation and management of waste. Pollution prevention programs at the sites are instrumental in achieving cost reductions for individual projects. The financial benefits of pollution prevention typically extend beyond the avoided costs of waste management and often accrue to a number of organizations at a given site.

Contract Reform. The largest portion of annual EM program funds is allocated to contractors that execute the work that accomplishes the cleanup mission. Reforms in contracting mechanisms offer the potential for significant savings. The EM program is developing site-specific contract strategies to improve overall program efficiency. Specific elements of these strategies include:

- Increased use of contractor incentives for improved performance (quality results and accelerated completion) and disincentives for poor performance;
- Additional privatization of certain EM cleanup activities by encouraging free market principles through a more open, competitive bidding process;
- Increased use of performance-based contracting mechanisms (for example, competitively awarded fixed-price contracts) to encourage more efficient cleanup; and
- Additional focus on linking work planning to the way contract types are selected, the incentives, and the make or buy process.

To ensure that sites work to implement the strategies, EM has undertaken a review of current contracting practices, focusing on integration of related activities and the periodic sharing of lessons learned to identify the contract vehicles most likely to facilitate the completion of the work. In addition, EM requested that sites report both quantitative and qualitative improvements in implementation of performance-based management contracts and the increases in dollar value or numbers of competitively awarded fixed-price contracts, including privatization contracts.

The improvements described above are being implemented at sites at which accelerated completion of the site scope of work is planned. Sites currently funded under the Closure Account have adopted new contracting principles that provide both incentives for accelerating cleanup and meaningful disincentives for falling behind schedule. Such a dual approach is crucial to the overall goal of making accelerated completion a reality. Eventually, each of the sites funded under the Closure Account will reach a stage at which the site managers can quantify required completion activities fully and award a competitive, performance-based contract, much like the contract awarded recently at the Miamisburg Environmental Management Project in Ohio.

Lessons Learned. As organizations perform the same activities repeatedly, they learn to do them more efficiently. Therefore, the cost (in constant dollars) of performing such activities declines. Data prepared by the Bureau of Labor Statistics, which measures productivity in the U.S. economy, indicate that, in the manufacturing sector of the economy, productivity has increased at an average annual rate of approximately 2.5 percent for the past 25 years. Therefore, in the average manufacturing industry, the cost of performing an activity is reduced by approximately one-half every 25 years. Although the EM program includes numerous technically complex, one-of-a-kind challenges and may not be able to match the industrial sector as a whole, there nevertheless are significant opportunities to improve productivity (that is, to achieve enhanced performance).

The EM program is an active participant in DOE's Lessons Learned program, a multifaceted initiative that uses information technologies to link Lessons Learned programs; rapidly transfer time-critical information about lessons learned to key points of contact; report upcoming events, such as conferences; and provide access to pertinent information available from sources outside DOE.

In addition, the EM program is reviewing PBSs to identify cases in which sharing of lessons learned might provide cost savings. For example, in deactivation and decommissioning of facilities, some sites are conducting smaller-scale projects during the period from 1998 to 2006, while other sites are conducting major deactivation and decommissioning work from 2020 to 2040. If the EM program can capitalize on lessons learned during the early years, significant savings may be achievable for later projects.

4.2.2 *Implementing Enhanced Performance*

The *Discussion Draft* identified cost reduction targets to eliminate differences between baselines and assumed funding levels entirely through enhanced performance. Initially, the targets in the *Discussion Draft* were estimates based on the experiences of DOE, organizations in the private sector, and other government agencies. These targets were based on assumptions that the EM program would:

- Reduce support costs to 30 percent of site costs by FY 2000;
- Achieve annual productivity improvements of 3.5 percent for definable (or pure) projects; and
- Achieve annual productivity improvements of 6 percent for operations (or operational projects).

Many reviewers of the *Discussion Draft*, however, questioned the validity of cost estimates based on these assumptions because they were derived from "across the board" application of the assumptions rather than by modifying specific project scope, schedule, and costs in the site baselines. The Environmental Management program has taken this reviewer criticism to heart; as a result, life-cycle cost estimates of the cleanup program are derived entirely from the sites' baselines in *Paths to Closure*. Thus the only enhanced performance reflected in the life-cycle cost estimates in *Paths to Closure* are those documented in site baselines.

However, EM is still pursuing the strategy of accelerating cleanup and reducing costs. Using the above assumptions in the *Discussion Draft* as a starting point, EM conducted a series of "workouts" with several sites. The objectives of the workout sessions were to identify opportunities to reduce costs significantly, increase efficiency, define better ways of managing resources and environmental objectives, and incorporate the resulting savings in site baselines. During the summer and fall of 1997, EM sponsored workouts at the Hanford Site, Idaho National Engineering and Environmental Laboratory, Rocky Flats Environmental Technology Site, Carlsbad Area Office, and the Savannah River Site. This round of workouts focused on performance enhancement targets and actions necessary to achieve those targets.

By using the workout process, Field Office Managers and contractors committed to enhanced performance goals for FY 1998 and FY 1999. FY 1998 and FY 1999 were a focus for two reasons: (1) the need to ensure full compliance in these years and (2) the goal of maximizing savings in the short term for reinvestment in the following years. The workout sessions achieved the results illustrated in Exhibit 4-3.

Sites have stated that since the targets were identified, total baseline costs have been reduced by over $5.6 billion based on identified opportunities to reduce cost and become more efficient. Unfortunately, during this same time, some sites have incurred some work scope growth, which offsets the substantial gains made by these performance enhancement opportunities. To help further lower costs, sites have targeted an additional $2.5 billion in enhanced performance savings. Despite these most recent targets, sites must still strive for additional enhanced performances; committing to additional enhanced performances will allow additional work scope to be completed for the same amount of money with resulting acceleration of site completion dates.

The Environmental Management program is deferring the establishment of accelerated closure dates and reduced life-cycle costs for most sites based on stakeholder concerns. After analysis of existing data, EM can establish credible acceleration goals based on the likelihood and difficulty of achieving technology development, integration, and other enhanced performance opportunities. EM plans to establish these acceleration goals in the 1999 update to *Paths to Closure.*

Exhibit 4-3

Summary of Site Workout Results

Office	Areas of Attention to Achieve Savings	FY 1998-99 Savings
Richland	The site is reducing direct/support areas, streamlining redundancy areas with contractors, maximizing use of contracting incentives, and exerting greater effort in implementation of new technologies.	$475 million
Savannah River	The site is deferring some work to accelerate "mortgage" reducing projects, reducing overlapping contractor responsibilities, using manpower more effectively, re-engineering processes to simplify the work needed to complete a task, and collaborating with regulators for scope changes on environmental restoration activities and safeguards and securities programs.	$300 million
Carlsbad	The site is working to ensure that it opens on schedule and is able to receive wastes from other sites as scheduled. By continuing to work to meet this milestone, savings will presumably result from other sites who are disposing the waste. In addition, Carlsbad has been able to achieve past efficiencies from expediting some activities.	$12 million
Idaho	EM and the site discussed several options to achieve further efficiencies during the workout but none appeared able to produce significant results. The site has a system in place that produced past improvements on various projects, allowing acceleration on other projects. Nevertheless, Idaho agreed to re-examine areas of efficiencies where future savings might be possible.	$52 million
Rocky Flats	The site goal is to accelerate site completion activities to 2006.	[a]

[a]Twelve percent per year positive schedule variance against the life-cycle baseline

4.2.3 *Requesting Additional Funds*

The budget is determined through an annual budget process (see text box). EM works with the Department and the Administration to request sufficient funds for compliance, consistent with its continued commitment to compliance. EM's needs are weighed during the budget process against other DOE and federal government priorities and the amount appropriated to EM has typically been less than the full request. Therefore, while EM could conceivably eliminate the difference between planning and funding levels by receiving more funding, fiscal realities are such that closing the gap completely by this mechanism is unlikely.

Process for Determining EM's Budget

- EM requests sufficient funds to comply with applicable environmental requirements as directed by Executive Order 12088.
- EM also requests funding for
  - "Minimum-safe" activities (DNFSB recommendations and to protect worker safety and health);
  - High-priority activities for the management and closure of sites; and
  - National programs and federal oversight at a level necessary and sufficient for EM.
- The Department works with the Administration to formulate a budget, balancing Department and other federal priorities.
- The President transmits a budget to Congress, which passes appropriations legislation.
- After Congress appropriates funds to specific accounts (Closure, Project Completion, Post-2006 Completion), the Department allocates each account to sites. See Section 5.4 for a description of each account.

4.2.4 *Meeting Immediate Health and Safety Needs*

If performance enhancements are not sufficient to address funding differences—either real or projected—at specific sites and additional funding requests are not successful, EM plans to pursue several options. In cases where new work is required immediately to protect safety and health, and related costs exceed available appropriations, the Department will shift funds from lower priority activities to ensure that public health and safety are adequately protected. The Environmental Management program will work with stakeholders, regulators, and Tribal Nations to address site priorities and proposed work deferrals, and will seek the reprogramming of any funds that may be necessary.

4.2.5 *Addressing Small, Projected Funding Differences*

Where performance enhancements are insufficient and small funding differences are projected at some sites in budget "outyears" (as is the case in FY 1999), the Environmental Management program will work with stakeholders, regulators, and Tribal Nations to identify funding for activities not required to maintain compliance or other high priorities to address such differences.

4.2.6 Addressing Larger Funding Differences in the Future

In future years where larger funding differences are projected, the Department intends to work with the Office of Management and Budget to seek additional funds for vitally important missions. Also, through site acceleration, it is DOE's goal to make additional resources available in the "outyears." DOE will propose shifting these resources from completed sites to other sites. No matter how successful these efforts are, however, the discipline of working within binding budget ceilings means that the Environmental Management program must engage in an active dialogue with stakeholders, regulators, and Tribal Nations about activities and programs at each of the Department's sites—and collectively make hard choices regarding priorities. The Environmental Management program will seek adequate funding to meet safety requirements and compliance obligations—but also will attempt to do more under limited funding projections.

The Environmental Management program is committed, therefore, to work with stakeholders, regulators, and Tribal Nations to review all aspects of the Department's environmental programs, including activities covered in enforceable agreements and activities that are not required under those agreements, to reach agreement on site programs that balance many competing priorities and needs. The Environmental Management program expects the strategy and the review of program options embodied in the development of *Paths to Closure* to become an important element of this effort.

Chapter 5

A Management System to Support the EM Program

To support the conceptual goals of accelerated cleanup and cost savings presented in *Paths to Closure,* the Office of Environmental Management (EM) has developed a new management system that consolidates planning, budgeting, and management functions. The new system, the Integrated Planning, Accountability, and Budgeting System (IPABS), makes a series of fundamental changes and improvements in EM's business processes. For the first time, EM will use a single framework for all its activities, linking planning, performance measurement, and the budget formulation and execution processes. This chapter presents the major components and processes of IPABS, which will support implementation of EM cleanup program:

- Baseline Management
- Program Management Tools
- Performance Measurement
- Budget Formulation
- Management Initiatives
- Program Evaluation

Exhibit 5-1 below presents a side-by-side comparison of the most significant changes in the EM program management system. The sections that follow

Exhibit 5-1

Fundamental Changes in EM Management Through IPABS

Former Process	IPABS Process
Activity-based	Project-based
Multiple database systems	One integrated set of corporate data
Multiple large data calls each year	Single large annual data call (with smaller updates as necessary)
Three year budget focus	Life-cycle focus integrated with three-year budget window
Overlap between Headquarters and Field management roles	Field focus on project management. Headquarters focus on policy, planning, integration, high-visibility projects, and programmatic risk mitigation

present more detailed discussions of IPABS advancements in each of the areas described above.

EM developed the changes and improvements in the management system in the context of the cleanup program. Consequently, EM considered the implications of each change on all aspects of its business processes. The final IPABS vision represents an integrated process, resulting in improved efficiency. Exhibit 5-2 presents a summary view of the IPABS process.

Exhibit 5-2
Diagram of IPABS Management System

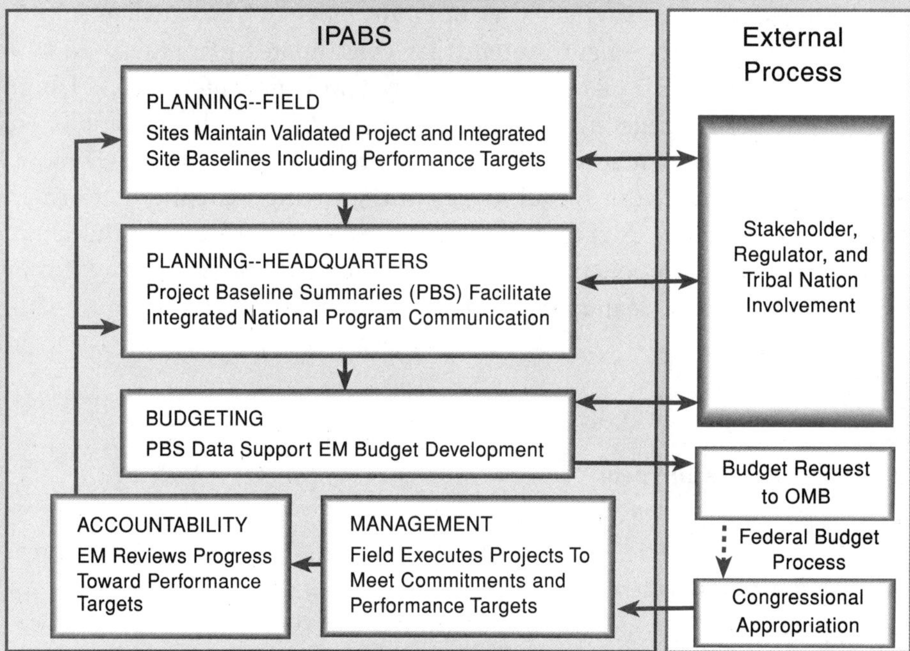

5.1 Baseline Management

A key element of IPABS is the baseline management framework that organizes the scope, schedule, and cost of all future cleanup activities into discrete projects. Historically, during the nuclear weapons development and production phase, sites used level-of-effort management approaches. In contrast, site baselines, built from individual project baselines, are the foundation for *Paths to Closure*. The focus on projects will enable more effective Field management, resulting in greater cost savings and accelerated completion. In addition, EM has established a change management process to track changes to the project structure and to maintain a consistent focus on achieving enhanced performance goals.

5.1.1 *Integrated Site Baselines*

The overall EM management strategy begins with the development of site baselines. Sites are responsible for developing detailed project baselines for all field projects, consisting of activities conducted in the EM program (e.g., environmental restoration, waste management, infrastructure, and long-term surveillance and monitoring). Each project must have a defined scope that guides managers in implementing each step of the cleanup. In addition, each project includes a quantitative expression of the engineering approach (i.e., scope, technical approach, schedule, cost requirements, and uncertainties) against which the status of resources and the progress of the project can be measured. All EM projects at a site comprise the integrated site baseline. Site baselines span the life cycle of all projects at the site and present a clear definition of overall cleanup requirements, individual cleanup milestones, critical interactions between projects, and costs over time.

5.1.2 *Baseline Validation and Change Control*

Once a site develops its integrated baseline, it is responsible for validating and maintaining it to reflect the most current state of planning at the site. The objective of baseline validation is to ensure that the baseline is defensible relative to scope, schedule, and cost. A credible and independent validation of each site's baseline is an expectation of Congress, the Office of Management and Budget (OMB), local stakeholders, and Tribal Nations.

A site must also reflect any changes to its planning baseline in its integrated baseline. EM has developed the outline for a disciplined change control process to manage and document changes to site baselines. A detailed process is under development. The process addresses three types of change that represent different levels of impact to the EM program (see Exhibit 5-3). Depending upon the type of change, different change control procedures are required. This tiered approach allows the sites freedom to manage their baselines efficiently, while enabling Headquarters to review changes that affect the entire program.

Exhibit 5-3

Levels of Change in EM Baseline Change Control Process

Change Type	Description	Requires HQ Approval
1 EM Policy Decisions	Policy decisions affecting the entire EM program or multiple sites	Yes
2 Major Baseline Adjustments	Changes to project end states, end dates, milestones on high-visibility projects, and changes that affect multiple sites	Yes
3 Limited Baseline Adjustments	Limited changes affecting a single project's or site's scope, cost, or schedule	No

5.1.3 *Relationship of Baseline Changes to the Annual* Paths to Closure *Report and DOE's Annual Financial Statement*

The EM program expects sites to change their baselines as necessary to reflect the most current state of planning as discussed in the previous section. Although site baselines will change as necessary, the Environmental Management program plans to publish updates to *Paths to Closure* each year. In addition, the Department publishes an annual financial statement in March reflecting its financial status as of the end of the fiscal year ending the previous September. This section discusses how EM plans to manage the relationship between continuously changing site baselines, annual *Paths to Closure* updates, and annual Department financial statements.

The relationship between changing site baselines and annual *Paths to Closure* updates is relatively straightforward. Sites should make changes to baselines as necessary, independent of *Paths to Closure* updates. Each year, sites will be asked to review and revise their baselines as part of the annual *Paths to Closure* update.

The relationship between changing site baselines, *Paths to Closure* updates, and the Department's annual financial statement is more complex. The complexity arises because sites may change baselines in between publication of the annual *Paths to Closure* update and the end of the fiscal year in September. Thus the Department's financial statement, which should reflect the Department's financial status as accurately as possible as of the end of the fiscal year may not agree with the published *Paths to Closure* update for that year.

The decision rule for incorporating baseline changes made after publication of the annual *Paths to Closure* update into the financial statement will focus on whether or not sites have formally approved baseline changes. Formally approved changes as of September 30 will be incorporated into the Department's financial statement. Changes not formally approved will be evaluated for possible incorporation into the Department's financial statement. For sites with formal change control systems, formally approved means that the change has been approved under the system. For sites with no system, formally approved means that site senior management has approved the change. Exhibit 5-4 illustrates how annual *Paths to Closure* report costs will be modified to accommodate annual financial statement needs.

Exhibit 5-4

Changing *Paths to Closure* Report Costs to Support Department Annual Financial Statement Needs

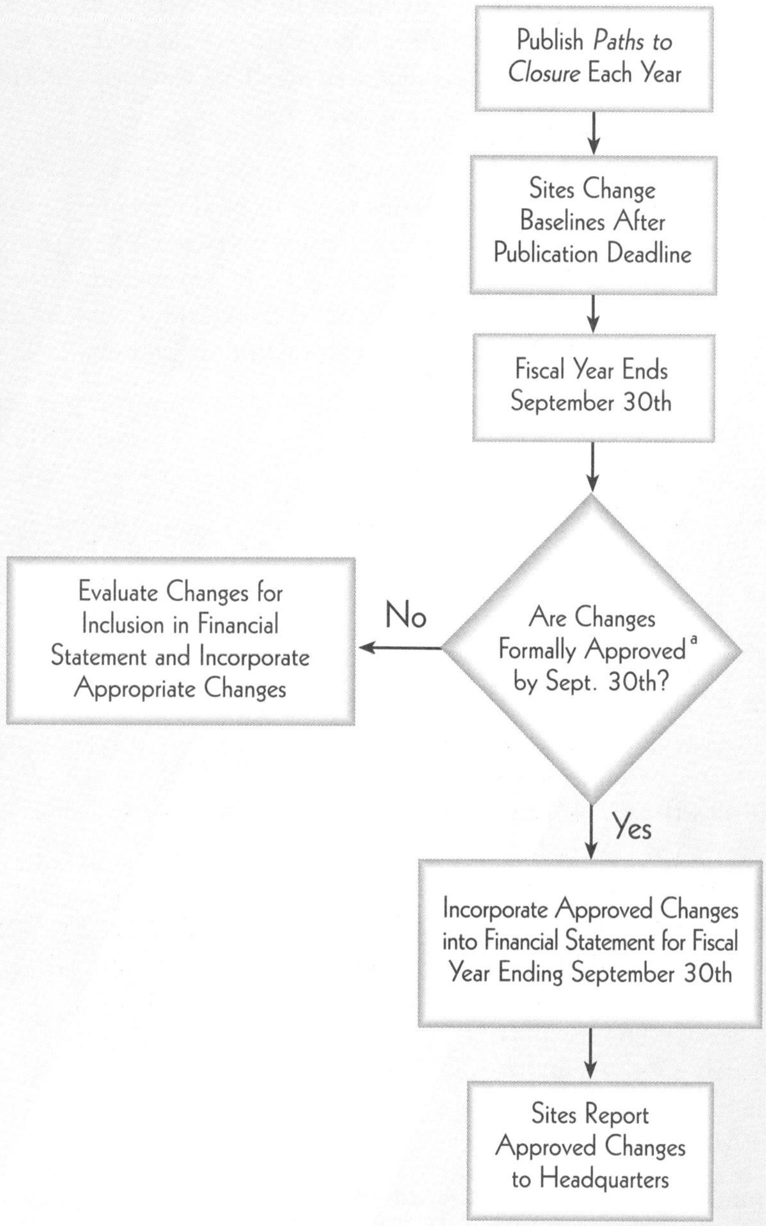

[a] For sites with formal change control systems, formally approved means that the change has been approved under the system. For sites with no system, formally approved means that site senior management has approved the change.

5.2 Program Management Tools

The integrated site baselines are the basis for a consolidated planning and program management capability in the Field and at Headquarters. EM will maintain summary level information on all Field projects and site planning information in a single database. EM will update this data primarily through a single annual data call, replacing the multiple, disconnected data calls required to support the previous data management systems. This database will enable EM to maintain more consistent information over time.

The data revolve around a set of management tools: 1) Project Baseline Summaries (PBSs), 2) Waste/Material disposition maps, 3) Critical Closure Paths, and 4) Programmatic Risk Assessments. Together, these tools enable EM to plan, budget, and execute work more effectively. They also allow EM to focus management attention on projects critical to the completion of the cleanup mission and direct technology development efforts to support those critical projects.

5.2.1 Project Baseline Summaries

Field projects that have common attributes, such as a common assumed end state, geographic location, or activity type are typically organized into IPABS projects (see text box). The individual Field projects which make up integrated site baselines are organized into IPABS projects for purposes of planning, budgeting, and management at the complex-wide level (see Exhibit 5-5). The Project Baseline Summary (PBS) is the single, summary-level report that describes the major management characteristics of each IPABS project.

Projects: Building Blocks of the EM Program

All EM projects must have:

- Logically organized components
- A defined start and end date
- A defined end state
- A reasonable size
- Milestones that demonstrate interim progress
- A validated baseline (cost, scope, schedule)
- Performance measures
- A designated DOE Project Manager

The PBS functions as the main source of project information at the Headquarters level and includes the scope, schedule, cost, life-cycle performance measurement metrics and annual performance targets, financial history and budget, and other information such as risk and assumptions. PBSs maintain data at a summary level to facilitate planning and program management at the national level, and they are directly linked to the more detailed project baselines developed at the site level. Summary level PBS data will be used for budget formulation and project performance tracking.

Exhibit 5-5
Integrated Baseline Management Framework

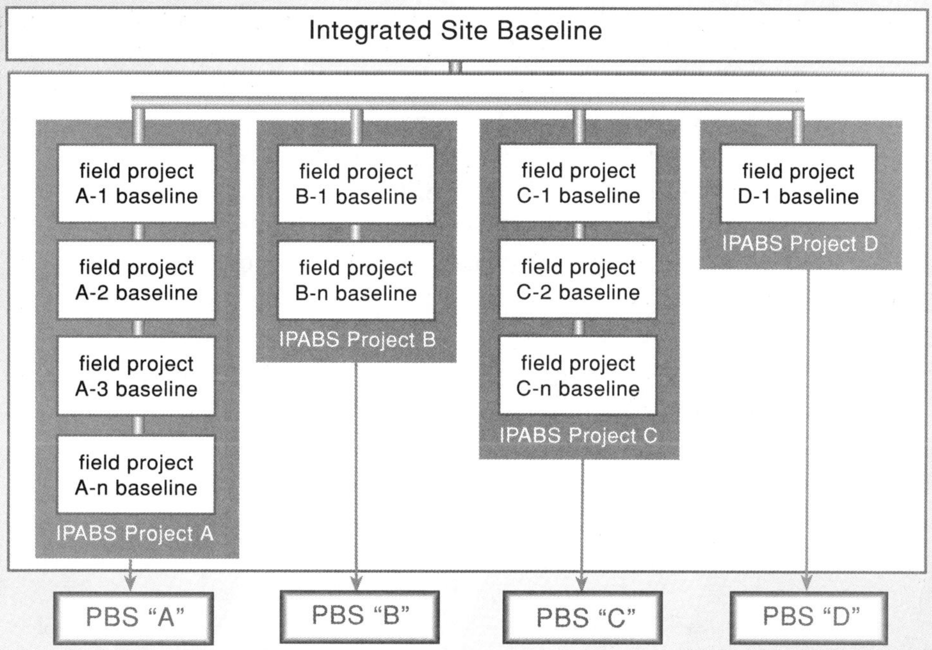

5.2.2 Waste and Material Disposition Maps

Waste and material disposition maps are graphical representations of each site's current conceptual approach to managing wastes, nuclear materials, and contaminated media from current status through storage, treatment, and disposal on- and off- site. These maps will provide stakeholders, regulators, and Tribal Nations a clear understanding of waste and materials disposition paths that have been decided, and current planning assumptions in cases where decisions have not yet been made and will enable more meaningful stakeholder participation in national planning efforts.

5.2.3 Critical Closure Paths

Site *Paths to Closure* reports describe "critical closure paths" for the major activities required for site closure. The critical closure path is a streamlined schedule of high-level activities, events, and/or decisions that warrant management attention and must occur "on schedule" to achieve the planned site closure date. These paths identify the set of activities that govern overall completion of EM scope at a site, including critical milestones and interdependent projects.

5.2.4 *Programmatic Risk Assessments*

To provide a means to elevate key issues and focus management attention, sites have identified those activities and events (key interim milestones) that must occur if the EM program is to remain on schedule and correspondingly within cost. For each such activity or event, sites have assigned a programmatic "risk" score in each of three areas: technology (do we have the technology to do our work?), scope (do we know how much work there is to do?), and intersite (do we know how and where we plan to store, treat, and dispose of material and waste?). These risk estimates will help EM prioritize funding among critical projects across the complex and identify areas requiring increased management attention and planning effort. Appendix D contains more information about programmatic risk.

5.3 Performance Measurement

EM has developed a single set of corporate performance metrics that focus the organization on achieving the goals and objectives identified in the *Paths to Closure* report, as well as on those crosscutting areas essential to accomplishing program results effectively and efficiently (i.e., financial, safety and health, risk reduction, and stakeholder trust and confidence measures). Tracking these metrics will help EM assess the outcomes of key activities as compared to planned goals, determine progress towards achieving the projects' and sites' assumed end states, and improve program performance at all organizational levels. In addition, measuring and tracking performance provides Congress and OMB with data to perform their oversight responsibilities.

EM Performance Metrics

EM's performance metrics data will be collected at a number of levels and will reflect key objectives of EM activities and crosscutting issues:

- Waste stored/treated disposed
- Release sites completed
- Facilities deactivated/decommissioned
- Material stabilized/made disposition-ready
- Technology deployment
- Risk reduction
- Safety and health
- Land released to public
- Pollution prevention
- Stakeholder trust and confidence

Performance metrics provide the link between the processes of planning, budgeting, executing, and evaluating. As such, performance measurement is a key component of all aspects of IPABS:

- **Planning.** As an integral part of the *planning* process, each site establishes performance goals against EM's corporate measures as applicable to their work scope. Planning information will inform the budget process.

- **Budget Formulation.** During the *budget formulation* process, performance information will be used to justify and defend EM's budget to OMB, Congress, stakeholders, regulators, and Tribal Nations. Performance goals that were established during the planning phase may be adjusted to reflect the results of congressional actions and field office baseline changes, as necessary.

- **Budget Execution.** Site project managers and contractors will *execute* their work scope in accordance with the approved work plans.

- **Program Evaluation.** Program results will subsequently be *evaluated* against the pre-established site and project performance measures goals and will be reported as part of the Assistant Secretary's Quarterly Management Reviews.

5.4 Budget Formulation

Each year, EM formulates a budget to satisfy DOE, OMB, and congressional mandates. While *Paths to Closure* is not a budget document, it is intended to inform budget formulation by establishing an overall strategic plan. Consistent with the 2006 vision, the budget is formulated by assigning projects to the following three program accounts:

- **Closure** includes all projects at sites closed by 2006 without a continuing DOE mission.

- **Project Completion** includes sites completed by 2006 with an ongoing DOE mission, and projects completed by 2006 at sites with cleanup work continuing after 2006.

- **Post-2006 Completion** includes projects that are expected to require work beyond FY 2006.

The new structure identifies three additional accounts: Technology Development, Program Direction (i.e., federal salaries), and Privatization projects. These six accounts are designed to allow Field managers more flexibility in using their funding more effectively to meet programmatic goals.

In keeping with the IPABS commitment to integrating planning, budgeting, and management functions, each project is assigned to one of these new budget accounts. As such, the budget process will be directly related to the cost estimates and performance metrics maintained in the Project Baseline Summaries. This will enable EM to develop more effective budgeting strategies that respond to progress in the Field and allocate appropriate funding to meet goals as expressed in critical closure paths and programmatic risk assessments.

5.5 Management Initiatives

IPABS not only integrates and streamlines EM's planning and budgeting process, but also improves the execution and management of EM activities. Three new management initiatives comprise the IPABS management reform efforts: 1) clarifying the Field and Headquarters management responsibilities, 2) elevating personal accountability through management commitments between Headquarters and Field Managers, and 3) instilling new incentives for enhanced performance and project acceleration through contract reform.

5.5.1 Clarifying Management Responsibilities

IPABS shifts the focus on management and execution of projects to the Field where the work toward closure is accomplished. The overall strategy for managing the Closure Account projects is for the Field to manage the planning, programming, and execution of its projects. Headquarters will work with the sites in preparing cleanup plans and, in partnership with the site, will assist in achieving cleanup objectives.

EM Headquarters has many roles for providing assistance to the Field. In its role of site advocacy, Headquarters personnel are responsible for working within the Department, OMB, and the Congress to obtain appropriate budget levels. Headquarters develops and implements cross-complex solutions for material consolidation and waste treatment, storage, and disposal. Headquarters also establishes necessary policies for the effective execution of cleanups. EM Headquarters staff serve as facilitators across Department Headquarters Offices and other agencies to assist the sites with meeting their performance commitments. Finally, Headquarters coordinates with stakeholders at a national level.

The Operations/Field Offices are responsible for awarding contracts, overseeing contractors, and the assurance of the health and safety of workers. Other responsibilities include developing project structure and definition; establishing project baselines; conducting performance assessments; and working with elected officials, federal/state/local regulators, Tribal Nations, other governmental agencies, stakeholders, and the public to implement the EM cleanup program at their sites.

5.5.2 Establishing Management Commitments

To establish more personal accountability for cleanup progress, the Assistant Secretary for Environmental Management and each Site Manager sign agreements for the execution year that commit each site to accomplishing a certain scope of work. These commitments are discrete examples of the focus on field-level responsibility and accountability for cleanup accomplishments. EM tailors these commitments to individual Operations/Field Offices and will provide a balanced approach to determining critical program expectations and for assessing EM's progress towards meeting key programmatic and high visibility project goals and objectives.

5.5.3 Improving Contract Management

EM's management system includes a range of improvements in the writing and execution of contracts. These improvements will ensure that EM contracting practices are consistent with the cost-effective achievement of *Paths to Closure* goals. IPABS envisions four specific contracting improvements:

- Increased use of contractor incentives for improved performance (e.g., quality results, accelerated completion) and disincentives for poor performance;
- Additional privatization of certain EM cleanup activities by encouraging free market principles through a more open, competitive bidding process;
- Increased use of performance-based contracting mechanisms (for example, competitively awarded fixed price contracts) to encourage more efficient cleanup; and
- Additional focus on linking work planning and the way contract types are selected, the incentives, and the make or buy process.

To ensure that all EM sites work towards implementing these strategies, EM has undertaken a review of current contracting practices, focusing on integration of related activities and the periodic sharing of lessons learned to determine the most favorable contract vehicles for accomplishing EM work. In addition, EM requested sites to report on the quantitative and qualitative improvements in their implementation of performance-based management contracts and the increases in dollar value or numbers of competitively awarded fixed price contracts, including privatization contracts.

These improvements are underway at sites planning on accelerated site work scope completion. Sites currently funded out of the Closure Account have adopted new contracting principles that provide incentives for accelerating cleanup and disincentives for falling behind schedule. This dual approach is crucial to the overall goal of making accelerated completion a reality. Eventually, each of the Closure Account sites will reach a stage when the site managers can fully quantify required closure activities and award a competitive, performance-based contract, much like the recent contract at the Miamisburg Environmental Management Project in Ohio.

5.6 Program Evaluation

Each of the components of IPABS described above enables EM to conduct a thorough evaluation of annual cleanup progress at the end of each fiscal year. Performance metric data can be summarized and compared against management commitments and enhanced performance goals. EM can use programmatic risk and critical closure path data to focus their performance reviews on PBSs critical to the completion of the EM program. Beginning with the 1999 update of *Paths to Closure*, EM plans to conduct a thorough evaluation of cleanup progress achieved during FY 1998 and report on that progress. Baselines in the current *Paths to Closure* will serve as the basis against which progress will be measured.

Chapter 6

Stakeholder, Regulator, and Tribal Nation Involvement

In developing and implementing its cleanup program, the Environmental Management program (EM), at both Headquarters and at sites, has placed a high priority on receiving input from all interested parties and incorporating revisions in response to those views into the site cleanup strategies as their development proceeds. However, responding to the variety of concerns continues to be a challenge. Congress, Tribal Nations, state and local governments, regulatory agencies, workers, environmental groups, citizen groups and advisory boards, the business community, academic institutions, and individuals all have unique perspectives and roles in the formulation of site cleanup strategies. In responding to input and feedback, the EM program has hoped to develop site strategies that fairly balance diverse and sometimes conflicting perspectives.

The June 1997 National and Site versions of *Accelerating Cleanup: Focus on 2006 Discussion Draft* were developed with the intent to identify the concerns of stakeholders, regulators, and Tribal Nations. The December 1997 *Preliminary Responses to Comments* document initially responded to the noted concerns received during the *Discussion Draft* comment period and formed the basis for continuing dialogue to further refine EM's cleanup program. Many of these concerns have since been addressed in *Paths to Closure*.

During the draft *Paths to Closure* 60-day comment period, which extended from publication in February 1998 until May 1, 1998, 39 sets of comments were received at Headquarters. EM identified over 260 individual comments on various facets of the report and grouped them into 13 categories: Relationship of *Paths to Closure* to Decision-making, Budget, Compliance, Contingencies, End States/ Stewardship, Safety and Health, Data Quality, Waste and Materials Disposition, Transportation, Enhanced Performance, Privatization, Technology Development, and Public Participation.

The following subsections of this chapter discuss the comments received in these categories that are relevant to the cleanup program. EM intends to send out individual letters to respond to more specific comments not addressed in this chapter. Additional copies of this report can be obtained from the Center for Environmental Management Information at 1-800-736-3282.

6.1 Relationship of *Paths to Closure* to Decision-making

Many stakeholders and one Tribal Nation expressed concerns about the relationship between *Paths to Closure* and the processes EM uses to make specific cleanup decisions. In particular, stakeholders and the Tribal Nation are concerned that assumptions about site end states (i.e., planning end points), used to establish scope, schedule, and cost estimates for cleanup projects, will preclude their opportunities to participate meaningfully in the determination of ultimate end states for sites. In addition, several commentors expressed concern that EM did not have an integrated and stable management and cleanup approach. In response to these concerns, *Paths to Closure* contains a new section in Chapter 1 (see Section 1.3) that describes the relationship of *Paths to Closure* to EM's decision-making processes.

Addressing Stakeholder, Regulator, and Tribal Nation Comments

Comment Area	Addressed in Chapter
Relationship of *Paths to Closure* to Decision-making	1
Budget	2, 4, 5
Compliance	1, 4
Uncertainties/Contingencies	1, 4
End States/Stewardship	1, 3, E
Safety and Health	1, 4
Data Quality	5
Waste and Materials Disposition	1, 3, 5
Transportation	1
Enhanced Performance	4
Privatization	4
Technology Development	1, 4
Public Participation	6

Decisions in the EM program are driven by various statutory mandates, most notably the Comprehensive Environmental Response, Compensation, and Liability Act (CERCLA) and the Resource Conservation and Recovery Act (RCRA). Most decisions are made at the site level (with appropriate Headquarters oversight). Other decisions are made at the Headquarters level because of their complex-wide implications. In many cases, ultimate decision-making authority, in the sense of final approval authority, resides with EPA or state regulators.

Public participation is an important element of the EM program's decision-making process. The National Environmental Policy Act (NEPA) requires federal agencies to consider the environmental impacts of their proposed actions. NEPA also requires that the public be informed of, and have an opportunity to comment on, major federal actions significantly affecting the environment. Consistent with its obligations under NEPA, the EM program performs an appropriate level of environmental review in connection with its projects, with opportunities for public involvement. For projects managed under CERCLA, EM relies on the CERCLA process to incorporate NEPA values.

Paths to Closure outlines EM's current estimate of the scope, schedule, and cost for each site to complete the cleanup program. The estimate includes projects for which key site cleanup decisions have been made pursuant to CERCLA, RCRA, or other statutes, and projects where such decisions have yet to be made. Where decisions have not yet been made, sites make **assumptions** (e.g., site planning end states) about how those cleanup actions might be carried out so that sites can define work and develop schedule and cost estimates. In those cases where decisions have not yet been made, the Environmental Management program will follow the decision-making processes called for by the relevant statutory authority that governs the activity in question (e.g., CERCLA or RCRA) with appropriate environmental review.

Paths to Closure also includes cost estimates for federal salaries, investments in science and technology development, and miscellaneous support functions. EM sites and EM Headquarters make decisions through the budgetary process on the scope and pace of work for these activities.

Stakeholders and Tribal Nations will have significant opportunities to participate in all decision-making processes.

6.2 Budget

Based on a review of the draft *Paths to Closure*, stakeholders voiced a concern that the funding assumptions used to develop the document exceed current budget projections. As a result, stakeholders felt that current budget projections would not be sufficient to accomplish EM's cleanup mission as it is outlined in *Paths to Closure*. In addition, stakeholders noted that EM should be diligent in its efforts to request adequate funding to meet compliance agreements and maintain the safety and health of workers, the public, and the environment. Stakeholders also were concerned that EM seek stable funding for sites.

EM realizes the necessity of matching planning dollars with funding levels. *Paths to Closure* provides a funding guideline of $5.75 billion per year for the entire EM program, starting in FY 1999. This figure was set in October 1997, *prior to DOE receiving its FY 1999 and outyear budget targets from the President*. It was essential to establish a funding profile at that time in order to produce this report on schedule. In some cases, sites exceeded the $5.75 billion funding guideline in order to meet compliance commitments. Further discussion of EM's funding assumptions can be found in Chapter 4.

EM directs sites to request sufficient funding to meet applicable environmental requirements in accordance with Executive Order 12088. Specifically, during the annual budget process, EM asks sites to identify funding requirements to meet compliance agreements, court orders, settlement agreements, consent decrees, and federal, state, and local statutes and regulations. EM is continually working with the Office of Management and Budget (OMB) and Congress to demonstrate the need for adequate funding including sufficient resources to meet compliance needs. EM uses a systematic process to reduce overall life-cycle costs:

- Constantly seeking ways to enhance performance;
- Requesting additional funding and/or considering reallocation of funds among sites to address immediate health and safety needs;
- For small funding differences, using funding available for other EM programs at a site to address compliance-related project scope; and
- For larger funding differences, working with OMB to seek additional funds, and working with stakeholders, regulators, and Tribal Nations to review sites' environmental activities to reach agreement on site programs that balance many competing priorities and needs.

Chapter 4 presents an additional discussion of enhanced performance.

6.3 Compliance

In addition to the concern discussed in Section 6.2 that EM would not be able to meet its regulatory obligations given current budget projections, stakeholders expressed concerns that EM might sacrifice compliance or health and safety in order to achieve enhanced performance goals and accelerated cleanups and closures.

The first step in EM's budget formulation process is to identify the funds necessary for full compliance. Although reducing costs through productivity improvements continues to be pursued as a means of accelerating closures and maintaining compliance under lower funding scenarios, enhanced performance savings are only captured in site baselines once a clear plan for implementation has been developed. As stated before, EM will not sacrifice compliance to achieve enhanced performance or accelerated closure dates.

6.4 Uncertainties/Contingencies

The long-range planning and unique processes involved in cleaning up DOE sites necessarily involve reliance upon some assumptions. Many of the comments expressed a general concern that the key assumptions outlined in Chapter 1 and the uncertainty that they hold with respect to future cleanup activities are not being adequately accounted for in program planning. Stakeholders are concerned that EM is not conducting enough contingency planning with respect to major assumptions. Also, EM received many comments that there is uncertainty in cost and schedule estimates resulting from project-specific assumptions.

While detailed contingency plans have not been developed for all of the key assumptions, the potential impacts have been evaluated at a high level. At this time, EM has chosen to not expend the substantial resources that would be needed to develop detailed contingency plans given that the current assumptions appear reasonable. Implicit in this approach is the assumption that sites conduct appropriate contingency planning in the event that there is a funding shortfall.

With respect to project-specific assumptions, each site selected the level of contingency included in each project. Sites have used the best available information to develop cost and schedule estimates, and any future changes in planning assumptions (e.g., changes in scope, end state, cleanup approaches, etc.) will be reflected in future revisions to *Paths to Closure.* EM recognizes the variability with respect to contingency planning among and within projects. As baselines improve over time through validation efforts, greater consistency in contingency planning will be achieved. One method for identifying potential areas of uncertainty at the national level is the use of programmatic risk scores. The programmatic risk scores, as discussed in Chapter 3 and Appendix E, help to focus management attention on possible areas of uncertainty where further contingency planning may be warranted. In future versions of *Paths to Closure,* EM will consider the impact of safety on the programmatic risk score.

6.5 End States/Stewardship

Numerous comments were received from stakeholders and one Tribal Nation regarding EM's end state assumptions and the plans for sites once EM's cleanup mission is completed. Stakeholders viewed the inclusion of assumed end states in the draft *Paths to Closure* as a positive addition to each site's cleanup strategy. However, many of the comments reiterated a concern that end state assumptions have not been approved in accordance with regulatory requirements and stakeholder agreements. Other comments expressed concern over the lack of comprehensive plans or cost estimates for the long-term monitoring and stewardship that will be required at many of the sites subsequent to EM cleanup.

As discussed in Chapter 1, in Section 1.3, the defining of end states is an ongoing process. Establishing a *planning* end state allows the sites to develop a description of the work scope, cost estimates, and schedule for the site's cleanup. These assumed end states may or may not be the ultimate end states. EM maintains that current assumptions about end states do not preclude future change resulting from changes in site planning assumptions, improved technology, increased cost efficiencies, the availability of additional resources, and/or changes in stakeholder and Tribal Nation interests.

EM acknowledges the need for more comprehensive plans addressing its role at sites once the cleanup mission has been achieved. The initial focus had been on developing baselines to address the estimated costs associated with the major cleanup work scope such as environmental restoration, waste treatment/

storage/disposal, deactivation, decommissioning, and materials stabilization. With baselines now improving, an increased focus will be placed on assessing long-term stewardship needs and formulating plans for post-closure activities at the sites. Some sites have already developed these estimates which are currently reflected in their baselines. EM plans to continue its studies in this area, and provide estimates of costs and plans for long-term stewardship across the complex in the next version of *Paths to Closure*. A companion report to *Paths to Closure, Moving from Cleanup to Stewardship*, is also being developed to address the scope, schedule, and cost of DOE's stewardship activities. This report will aim to clarify cleanup goals and long-term stewardship intentions.

6.6 Safety and Health

Stakeholders have expressed concern that EM's emphasis, as reflected in the draft version of *Paths to Closure*, has shifted away from mitigating safety and health risks toward accelerating cleanup. Stakeholders fear that the safety and health of workers, the public, and the environment has been, or may be, compromised so that other goals such as enhanced performance may be accomplished.

EM remains committed to its policy to "Do Work Safely or Don't Do It!" and continues to include safety and health concerns as an integral part of project planning. In fact, the primary mission of the EM program is to reduce threats to safety and health posed by contamination and waste at DOE sites. The protection of workers, the public, and the environment is a factor included in the planning of each project. EM is a leader in Integrated Safety Management (ISM), an approach that incorporates safety and health concerns into project planning. Efforts will continue to focus on integration of the Department's overall ISM system and individual projects to ensure that cross-cutting facility and worker safety and health issues are addressed in a consistent and effective manner. Chapter 1 discusses the integration of safety and health throughout EM's program in greater detail.

EM does not view its goal of accelerated cleanup as being in conflict with its goal of maintaining safety and health standards. The philosophy behind *Paths to Closure* is to focus programmatic priorities on the safe, compliant acceleration of cleanup and site closure. EM will continue to seek productivity improvements, without jeopardizing health and safety standards.

6.7 Data Quality

EM received numerous comments from stakeholders who felt that the draft *Paths to Closure* had made significant strides in the extent and clarity of data presented. Stakeholders found the addition of the Conceptual Summary Disposition Maps and programmatic risk tables to be especially insightful. However, several stakeholder comments still expressed concerns over the quality of the data, noting inconsistencies and gaps in the level of detail provided.

EM has actively sought to improve the quality of data throughout *Paths to Closure.* The alignment of information presented in *Paths to Closure* with site baselines is a major step toward improving data quality. The iterative nature of the process has also led to improved data quality, and each subsequent update should be better. As an example of this effort to improve data quality, EM has improved the quality of the data contained in waste and material disposition maps. In order to mitigate data discrepancies in disposition maps, EM is taking an iterative approach to refine the information (see Section 6.8).

In conjunction with the evolution of *Paths to Closure*, EM has implemented a more comprehensive management system, the Integrated Planning, Accountability, and Budgeting System (IPABS). As further explained in Chapter 5, IPABS will integrally link the planning, accountability, and budgetary functions to achieve a higher degree of data quality and data consistency.

6.8 Waste and Materials Disposition

With respect to EM's waste and materials disposition data, many stakeholder comments focused on the newly added disposition maps. As mentioned above in the Data Quality section, most stakeholders viewed the disposition maps as a positive addition and made some suggestions for further refinements. However, many stakeholders expressed concern over the assumptions used in developing the disposition maps, especially with respect to intersite transfers. Several comments also advocated that plans for addressing newly-generated waste be developed and included in *Paths to Closure*.

Improving waste and materials disposition data was augmented in response to comments received on the *Focus on 2006: Discussion Draft.* EM developed a process of collecting data to communicate assumptions for managing waste and materials at each site in the complex. Based on the data collected, disposition maps were generated to reflect the current waste management assumptions at sites and to provide a look across sites. One clear benefit has been that disposition maps have catalyzed the necessary dialogue between sites regarding potential intersite transfers. By incorporating stakeholder comments and performing additional data collection, EM anticipates further refinement of waste and materials data leading to an even more effective tool for complex-wide communication, reporting, and analysis.

It is important to note, however, that disposition maps are not decision-making tools; they simply depict baseline planning assumptions. As decisions are made (through the processes described in Chapter 1) disposition maps will be refined to reflect any planning changes.

With respect to newly-generated waste, EM is assuming that generators will be financially responsible for managing and disposing of wastes appropriately. This transfer of responsibility has already been implemented at some sites and is expected to increase as FY 2000 approaches.

6.9 Transportation

Most of the comments received regarding transportation expressed a concern that EM has not fully developed and shared its transportation plans. Without comprehensive plans for the transport of waste, some stakeholders question the validity of assumed shipments discussed in the draft *Paths to Closure*. In addition, some stakeholders feel that transportation decisions have not given adequate weight to the risks involved in transporting certain types of waste.

EM recognizes the degree to which *Paths to Closure* relies on intersite transport of waste and materials to accomplish its goals. Although transportation issues have not been specifically addressed in this *Paths to Closure* report, they are an integral part of each site's decision-making process. A recently established Executive Steering Committee on Transportation is working to address transportation issues. In addition, EM has begun transportation systems engineering and anticipates providing more substantive information regarding complex-wide transportation in the 1999 version of *Paths to Closure*.

6.10 Enhanced Performance

Some stakeholders support EM's strategies to accelerate closures through enhanced performance, and advocate that EM continue to formulate strategies to achieve productivity improvements. Some stakeholders were nevertheless concerned that the adoption of enhanced performance techniques may lead to compromises in other facets of EM's cleanup mission in order for the underlying goals of acceleration and cost reductions to be achieved.

The enhanced performance savings reflected in baselines represent only those savings for which a feasible strategy has been adopted. EM views enhanced performance as a prudent management tool, and will continue to promote the development and employment of sound strategies to achieve productivity improvements. Chapter 4 provides a detailed discussion of EM's enhanced performance strategies and expectations.

6.11 Privatization

EM's promotion of privatization has been criticized due to a lack of data to support the hypothesis that enhanced performance will result from its employment. Many stakeholders questioned the merits of privatization which they claim has not been as successful in all cases as had been anticipated. Concerns were expressed that *Paths to Closure* continues to promote privatization despite evidence that it is not necessarily a means of reducing costs.

Currently, EM continues to support privatization strategies as a means to reduce risks and costs. Privatization as used in this context refers to a particular method of financing, contracting, and risk-sharing with the private sector for goods and services. In using privatization, EM is relying on market forces to set prices through competition for fixed-price contracts.

6.12 Technology Development

Many stakeholders see the potential for EM to enhance its performance through the adoption of new technologies, and encourage more investment in the development of feasible deployment strategies.

One of EM's enhanced performance strategies relies on the identification of areas where technological advancements would have the most beneficial impact on costs and schedule. The *Paths to Closure* process has identified projects and activities where new technologies have the most potential for reducing costs or accelerating schedules. With this information, EM will be able to target its resources for technology development where they will be most effective.

6.13 Public Participation

Some stakeholders feel that EM has addressed their comments and concerns in *Paths to Closure*. Yet, there remains room for more progress in carrying out EM's goals to incorporate stakeholder comments in the formulation of its cleanup program. Some stakeholders feel that certain areas of concern have not received appropriate response from EM. Other stakeholders feel that more opportunities for public involvement should be provided.

As discussed in Chapter 1, public participation is a crucial component in EM's successful completion of its cleanup program. Comments submitted are viewed as valuable feedback and guidance as the process of creating site strategies evolves into a sound cleanup program. EM has attempted to address most of the stakeholder comments received in response to the draft *Paths to Closure* document either through explicit changes incorporated in this version of *Paths to Closure* or in the discussion in this chapter. EM also plans to send to each commentor an individual letter, which will respond in greater detail to specific comments. The public's concerns will continue to be addressed in the ongoing development of the next version of *Paths to Closure*.

Many comments received were noted to be specific to the conditions at individual sites. Because each site has unique issues to resolve and decision-making occurs predominantly at the site level, most of these comments will be addressed in each site's *Paths to Closure* report.

Appendix A

List of Project Baseline Summaries

List of Project Baseline Summaries

Albuquerque Operations Office		
Albuquerque Operations Office	-	Albuquerque Miscellaneous Programs (WERC, HBCU, ITRD, NSUC, AIP-TX/MO)
	-	New Mexico Agreement in Principle (AIP)
Grand Junction Office (GJO)	-	GJO All Other Projects
Lovelace Respiratory Research Institute	-	Lovelace Respiratory Research Institute
Kansas City Plant (KCP)	-	Kansas City Plant Environmental Restoration
Los Alamos National Laboratory (LANL)	-	Nuclear Material Facility Stabilization Research and Development
	-	LANL Environmental Restoration
	-	LANL Waste Management - Newly Generated Waste
	-	LANL Waste Management - Legacy Waste
Maxey Flats	-	Maxey Flats Field Management Project
Monticello	-	Monticello Projects
Pantex Plant	-	Pantex Plant Site Remediation Project
	-	Pantex Waste Operations
Pinellas Plant	-	Pinellas Plant Closeout and Administration of Post-Employment Benefits
	-	Ground Water Cleanup (Pinellas Plant)
Sandia National Laboratories (SNL)	-	Sandia National Laboratories (SNL) Waste Management
	-	Sandia Environmental Restoration Project
South Valley	-	South Valley Superfund Site
Uranium Mill Tailings Remedial Actions (UMTRA) - Groundwater	-	UMTRA Groundwater
UMTRA-Surface	-	UMTRA - Surface Remedial Action Project

Carlsbad Area Office	
Waste Isolation Pilot Plant (WIPP)	- WIPP Base Operations - WIPP Disposal Phase Certification and Experimental Program - WIPP Transportation - WIPP Transuranic Waste Sites Integration and Preparation - WIPP Transuranic Waste Transportation Privatization

Chicago Operations Office	
Ames Laboratory	- Ames Remedial Actions - Ames Waste Operations
Argonne National Laboratory - East (ANL-E)	- ANL-E Program Management - ANL-E Decontamination and Decommissioning Actions - ANL-E Remedial Actions - ANL-E Waste Operations
Argonne National Laboratory - West (ANL-W)	- ANL-W Remedial Actions - ANL-W Waste Operations
Brookhaven National Laboratory (BNL)	- BNL Boneyard Waste - BNL Decontamination and Decommissioning Actions - BNL Program Management - BNL Remedial Actions - BNL Waste Operations
Chicago Operations Office	- Princeton Site A/B Payments - Site A Cleanup - Surveillance and Maintenance Activities - Chicago Operations Program Support
Fermi National Accelerator Laboratory (FNAL)	- Fermi National Accelerator Laboratory (FNAL) Waste Operations
Princeton Plasma Physics Laboratory (PPPL)	- Princeton Plasma Physics Laboratory (PPPL) Remedial Actions - PPPL Waste Operations

Headquarters / National Programs

Program Direction	- Program Direction
Technology Development	- National Risk Program
	- Environmental Management Science Program
	- National Science and Technology Development
Technical Support	- Technical Support to Environmental Restoration
	- Headquarters Program Integration
	- Environmental & Regulatory Analysis
	- Office of Waste Management
	- Support to Transition Activities
Other National Programs	- National Characterization Management Program
	- Emergency Preparedness Program
	- National Transportation Program
	- Packaging Certification
	- Pollution Prevention
	- Radioactive Source Recovery Program (RSRP)

Idaho Operations Office

Idaho Operations Office	- Science and Technology Coordination
Idaho National Engineering and Environmental Laboratory (INEEL)	- Low-level Waste/Mixed Low-level Waste Center of Excellence
	- Test Area North Remediation
	- Test Reactor Area Remediation
	- Idaho Chemical Processing Plant Remediation
	- Central Facilities Area (CFA) Remediation
	- Power Burst Facility/Auxiliary Reactor Area
	- Radioactive Waste Management Complex Remediation
	- Pit 9 Remediation
	- Sitewide Monitoring Area Remediation
	- Remediation Operations
	- Decontamination and Decommissioning
	- High-level Waste Pretreatment
	- High-level Waste Immobilization Facility (Privatized)
	- High-level Waste Treatment and Storage
	- Vitrified High-level Waste Storage
	- Low Activity Waste Treatment
	- Sitewide Landlord Operations
	- Idaho Chemical Processing Plant / Non-process Plant Operations

Idaho Operations Office (Continued)

Idaho National Engineering and Environmental Laboratory (INEEL) (Continued)	- INEEL Medical Facilities
	- INEEL Emergency Response Facilities
	- Security Facilities Consolidation Project
	- Electrical and Utility Systems Upgrade (EUSU) Project, Idaho Chemical Processing Plant (ICPP)
	- INEEL Electrical Distribution Upgrade
	- INEEL Road Rehabilitation
	- Health Physics Instrument Laboratory
	- Pre-FY 2007 Surplus Facility Deactivation Project
	- Post-FY 2006 Surplus Facility Deactivation Project
	- Pre-2007 INEEL Surveillance and Maintenance (S&M)
	- Post-2006 Surveillance, Maintenance, and Monitoring
	- National Spent Nuclear Fuel Program
	- Integrated Spent Nuclear Fuel (SNF) Program
	- Emptied SNF Facilities
	- Constructed New Facilities
	- Dry Transfer and Storage Project (Privatized)
	- INEEL Low-level Waste / Mixed Low-level Waste / Other Waste Program
	- National Low-level Waste Program
	- INEEL Transuranic Waste
	- Advanced Mixed Waste Treatment Plant (AMWTP) Asset Acquisition Project (Privatized)
	- AMWTP Production Operations
	- INEEL Sitewide Environmental Protection
	- Long-term Treatment/Storage/Disposal Operations
	- Integrated Waste Operations Program

Nevada Operations Office

Nevada Test Site (NTS)	- Program Integration
	- Agreements In Principle / Grants
	- Soils
	- Underground Test Area (UGTA)
	- Industrial Sites
	- Program Management
	- Transuranic Waste/Mixed Transuranic Waste
	- Mixed Low-level Waste
	- Low-level Waste
Nevada Offsite	- Off sites Remedial Action

Oak Ridge Operations Office

Oak Ridge Operations Office	-	Directed Support
Oak Ridge Reservation	-	Hazardous Waste Management
	-	Sanitary/Industrial Waste Management
	-	Mixed Low-level Waste Management
	-	Low-level Waste Management
	-	Transuranic Waste Management
	-	Transuranic Waste Privatization
	-	Y-12 East Fork Poplar Creek Remedial Action
	-	Y-12 Bear Creek Remedial Action
	-	Oak Ridge National Laboratory (ORNL) Melton Valley Watershed Remedial Action
	-	ORNL Melton Valley Watershed Deactivation & Decommissioning
	-	ORNL Bethel Valley Remedial Action
	-	ORNL Bethel Valley Deactivation & Decommissioning
	-	East Tennessee Technology Park (ETTP) Landlord
	-	ETTP Remedial Action
	-	ETTP Process Equipment Deactivation & Decommissioning
	-	ETTP Deactivation & Decommissioning
	-	ETTP Facility Safety Upgrades
	-	On-site Waste Management Facility
	-	Off-site Remedial Action
	-	Nuclear Materials and Facility Stabilization (NMFS)
Paducah Gaseous Diffusion Plant	-	Paducah Remedial Action
	-	Paducah Waste Management
Portsmouth Gaseous Diffusion Plant	-	Portsmouth Remedial Action
	-	Portsmouth Waste Management
Weldon Spring Site	-	Weldon Spring Disposal Facility
	-	Weldon Spring Waste Treatment
	-	Weldon Spring Long-term Surveillance and Maintenance

Oakland Operations Office

Energy Technology Engineering Center (ETEC)	-	ETEC Remediation
	-	ETEC Landlord
	-	ETEC Waste Management

Oakland Operations Office (Continued)

General Atomics	- Hot Cell Facility Deactivation & Decommissioning at General Atomics
General Electric	- General Electric Deactivation & Decommissioning (Environmental Restoration)
Geothermal Test Facility	- Soil Remediation at Geothermal Test Facility (GTF)
Lawrence Berkeley National Laboratory (LBNL)	- LBNL Legacy Waste - LBNL Newly-generated Wastes - LBNL Soils and Groundwater (Environmental Restoration) - LBNL Hazardous Waste Handling Facility Closure (Environmental Restoration)
Laboratory for Energy-Related Health Research (LEHR)	- LEHR Environmental Restoration - LEHR Waste Management
Lawrence Livermore National Laboratory (LLNL)	- Accelerated Waste Treatment - LLNL Main Site Remediation - LLNL - Site 300 Remedial Action - LLNL Base Program - LLNL General Plant Projects - LLNL Decontamination and Water Treatment Facility
Oakland Operations Office	- State Grants
Stanford Linear Accelerator Center	- Stanford Linear Accelerator Center (Environmental Restoration)
Separation Process Research Unit (SPRU)	- Separation Process Research Unit (SPRU)

Ohio Field Office

Ashtabula Environmental Management Project	- Ashtabula Remediation - Project Management, Site Services, Environmental Safety & Health
Columbus Environmental Management Laboratory (CEMP)	- King Avenue Site Decontamination - West Jefferson Site Decontamination - Project Management, Site Support & Maintenance

Ohio Field Office (Continued)

Fernald Environmental Management Program (FEMP)	- Facility Shutdown
	- Facility Deactivation & Decommissioning
	- On-site Disposal Facility
	- Aquifer Restoration
	- Waste Pits Remediation Project
	- Soils
	- Silos
	- Nuclear Materials
	- Thorium Overpack
	- Mixed Waste
	- Waste Management
	- Program Support & Oversight
Miamisburg Environmental Management Project (MEMP)	- Tritium Operations Transition
	- Main Hill Tritium
	- Legacy Waste
	- Main Hill Rad
	- Main Hill Non-rad
	- Special Materials / Plutonium Processing (SM/PP) Hill
	- Test Fire Valley
	- Soils
	- Facility Operations & Maintenance
	- Exit Support Project
West Valley Demonstration Program	- High-level Waste Vitrification and Tank Heel High Activity Waste Processing
	- Site Transition, Decommissioning, and Project Completion
	- Spent Nuclear Fuel
	- Project Management/Site Support

Richland Operations Office

Hanford	- 100 Area Remedial Action
	- 200 Area Remedial Action
	- 300 Area Remedial Action
	- Environmental Restoration Disposal Facility
	- Facility Surveillance & Maintenance - ADS 3500
	- Decontamination and Decommissioning
	- Post Closure Surveillance & Maintenance
	- Groundwater Management

Richland Operations Office (Continued)

Hanford (Continued)

- N Reactor Deactivation
- Program Management and Support
- HAMMER
- Mission Support
- B-Plant Sub-project
- Waste Encapsulation and Storage Facility (WESF) Sub-project
- Plutonium-Uranium Extraction Plant (PUREX) Sub-project
- 300 Area / Special Nuclear Materials (SNM) Sub-project
- Plutonium Finishing Plant (PFP) Deactivation
- PFP Stabilization
- PFP Vault Management
- 324/327 Facility Transition Project
- K Basin Deactivation
- Accelerated Deactivation
- Advanced Reactors Transition
- Transition Project Management
- Landlord Project
- Hanford Surplus Facility Program 300 Area Revitalization Project
- Tank Waste Characterization
- Tank Safety Issue Resolution Project
- Tank Farms Operations
- Retrieval Project
- Process Waste Support
- Process Waste Privatization Phase I
- Process Waste Privatization Phase II
- Process Waste Privatization Infrastructure
- Immobilized Tank Waste Storage & Disposal Project
- Tank Waste Remediation System Management Support
- Spent Nuclear Fuels Project
- Canister Storage Building Operations
- Solid Waste Storage and Disposal
- Solid Waste Treatment
- Liquid Effluents Project
- Analytical Services

Richland Operations Office

- Richland Directed Support
- Tank Waste Remediation System Regulatory Unit
- Pacific Northwest National Laboratory Waste Management

Rocky Flats Field Office

Rocky Flats Operations Office	-	Work for Others Project
Rocky Flats Environmental Technology Site (RFETS)	-	Buffer Zone Closure Project
	-	Waste Management Project
	-	Remediation Waste & Contingent Storage Project
	-	Special Nuclear Materials (SNM) Capital Support Project
	-	International Atomic Energy Agency (IAEA) Project
	-	SNM Consolidation Project
	-	New Plutonium Interim Storage Vault
	-	Plutonium Metals and Oxides Stabilization
	-	Plutonium Solid Residue Stabilization Project
	-	Plutonium Liquid Stabilization
	-	Uranium Disposition Project
	-	SNM Shipping Project
	-	Closure Caps Project
	-	Industrial Zone Closure Project
	-	Miscellaneous Production Zone Cluster Closure Project
	-	Building 371 Cluster Closure Project
	-	Building 707/750 Cluster Closure Project
	-	Building 771/774 Cluster Closure Project
	-	Building 776/777 Cluster Closure Project
	-	Building 881 Cluster Closure Project
	-	Building 991 Cluster Closure Project
	-	Building 779 Cluster Closure Project
	-	Utilities & Infrastructure Project
	-	Safeguards and Security Project
	-	Infrastructure Improvement/Replacement Project
	-	Analytical Services Project
	-	Rocky Flats Field Office - DOE Management
	-	K-H Project Management

Savannah River Operations Office

Savannah River Operations	-	DOE External Program Support
	-	DOE Program Support
Savannah River Site (SRS)	-	DOE Projects Line Item
	-	Wackenhut Services - Incorporated Savannah River Site Landlord Project

Savannah River Operations Office (Continued)	
Savannah River Site (SRS) (Continued)	- Savannah River Natural Resource Management & Research Institute - Ecology Lab Project - Flood Plain Swamp Project - Four Mile Branch Project - Lower Three Runs Project - Pen Branch Project - Steel Creek Project - Upper Three Runs Project - Program Management - Facility Disposition Program Planning - Heavy Water Components Test Reactor (HWCTR) Projects - 247-F Deactivation Project - F Canyon Deactivation Project - FB Line Deactivation Project - H Canyon Deactivation Project - HB Line Deactivation Project - 235-F Deactivation Project - Old HB Line Deactivation Project - P Reactor Deactivation Project - C Reactor Deactivation Project - R Reactor Deactivation Project - K Reactor Deactivation Project - L Reactor Deactivation Project - Receiving Basin for Off-site Fuels (RBOF) Deactivation Project - D Area Deactivation Project - M Area Deactivation Project - F Area Monitoring - H Area Monitoring and Minor Facility Monitoring - M Area Monitoring Project - D Area Monitoring Project - Reactors Monitoring Project - Heavy Water Storage Monitoring - RBOF Monitoring Project - H Tank Farm - F Tank Farm - Waste Removal Operations and Tank Closure - In Tank Precipitation (ITP) / Extended Sludge Processing (ESP) / Late Wash (LW) Operations - Vitrification - Glass Waste Storage

Savannah River Operations Office (Continued)

Savannah River Site (SRS) (Continued)

- Effluent Treatment Facility
- Saltstone
- Tank Farm Service Upgrades
- H Tank Farm Storm Water System Upgrades
- Tank Farm Support Services F Area
- High-level Waste System Upgrades
- Plantwide Fire Protection Line Item
- Operations Support Facility Line Item
- Plant Maintenance Line Item
- Domestic Water Line Item
- CFC HVAC Chiller Retrofit (96-D-471)
- Radio Trunking System Line Item
- Site Road Infrastructure Line Item
- High-level Drain Lines Line Item
- Health Physics Support Line Item
- Regulatory Monitoring and Bioassay Laboratory
- Infrastructure Line Item
- Operating Projects
- Decontamination of Laboratory Facilities, 772-F and 773-A
- F Area Stabilization Project
- H Area Stabilization Project
- Actinide Packaging Line Item
- Canyon Exhaust Line Item
- Neptunium (Np) Vitrification Line Item
- Nuclear Materials Storage
- Depleted Uranium Storage
- K Reactor Spent Nuclear Fuel Project
- L Reactor Spent Nuclear Fuel Project
- RBOF Spent Nuclear Fuel Project
- Heavy Water - D Area
- Alternate Technology Project
- Disassembly Basin Upgrade Line Item
- Spent Nuclear Fuel Transfer and Storage
- RBOF Process Support System Refurbishment
- Consolidated Incinerator Facility
- Transuranic Waste Project
- Mixed Low-level Waste Project
- Low-level Waste Project
- Hazardous Waste Project
- Sanitary Waste Project
- Pollution Prevention

Uranium Enrichment Deactivation and Decommissioning Fund / Uranium and Thorium Licensees

Reimbursement to Uranium and Thorium Licensees	-	Reimbursements to Uranium and Thorium Licensees under Title X of the Energy Policy Act of 1992
Uranium Enrichment	-	Contribution to the Uranium Enrichment Deactivation and Decommissioning Fund

Appendix B

Example Project Baseline Summary

Project Baseline Summary Report

Report ID Number: Q501

Data Version: **16-Jan-98** Print Date: **19-Feb-98**

Operations/Field Office: **Idaho** Site: **Idaho National Engineering and Environmental Laboratory**

HQ ID: **IDIN0570** Project: **INEEL Low-Level Waste/Mixed Low-Level Waste/ Other Waste Program (ID-WM-101)**

A.1. - Project Identification/Header Information

A.1.5. DOE Project Manager: Jeff T. Shadley

A.1.6. DOE Project Manager Phone Number:

A.1.7. DOE Project Manager Fax Number:

A.1.8. DOE Project Manager e-mail address:

A.1.14. Program Element: WM **A.1.15. Project Type:** Operational

A.1.9. Contractor Project Manager: M. C. Tieman

A.1.10. Contractor Project Manager Phone Number:

A.1.11. Contractor Project Manager Fax Number:

A.1.12. Contractor Project Manager e-mail address:

A.1.16. Is this a High Visibility Project (Y/N): No

A.2. Technical and Scope Narratives

A.2.1. Purpose of Project:

Predecessor Projects: ADS ID-4310-01, WROC Operations; ADS ID-4311-02, Low-Level Waste Operations; Portions of ADS ID-4302-01 FFCA Implementation and Waste TSD Optimization (Special Case Waste and off-site LLW disposal), ADS ID-4303-01 Waste Management General Plant Projects (for LLW and MLLW GPP projects), ADS ID-1001-01 High Level Waste (ICPP LLW handling, hazardous waste and MLLW storage plus industrial waste cuber operations)

The Idaho National Engineering and Environmental Laboratory (INEEL) has been supporting the Department of Energy (DOE) in nuclear energy research for over forty years. This research has routinely generated mixed low-level waste (MLLW), low-level waste (LLW), hazardous waste (HW), and industrial waste requiring treatment, storage, and/or disposal (TSD). Cost/benefit studies are routinely used to evaluate commercial treatment and disposal services, in lieu of INEEL services. Commercial facilities are used where they can be shown to be cost effective. The cost of treating other DOE site MLLW is included in this PBS. The only cost required to be paid by the other DOE sites include commercial disposal, if available at the time of treatment, and any required treatability studies.

The INEEL and other DOE sites generated and stored MLLW for years without having provisions for meeting the requirements of the Resource Conservation and Recovery Act (RCRA). The Federal Facility Compliance Act (FFC Act), passed in 1992, requires DOE to prepare a plan for the development of needed treatment capacity and technology for each facility at which DOE generates or stores mixed waste and hazardous waste. The INEEL has complied with the FFC Act and has an approved Site Treatment Plan (STP) and associated Consent Order. This project supports STP compliance by providing incineration, stabilization, macroencapsulation,

Project Baseline Summary Report

Report ID Number: Q501

Data Version: **16-Jan-98** Print Date: **19-Feb-98**

Operations/Field Office: **Idaho** Site: **Idaho National Engineering and Environmental Laboratory**

HQ ID: **IDIN0570** Project: **INEEL Low-Level Waste/Mixed Low-Level Waste/ Other Waste Program (ID-WM-101)**

A.2. Technical and Scope Narratives

sizing/sorting/segregation, and lead cask dismantlement services for the treatment of INEEL and other DOE Complex sites MLLW through FY2003 and commercial treatment of INEEL MLLW between FY2004 and FY2006 using the Advanced Mixed Waste Treatment Project (AMWTP).

DOE Order 5820.2A, Revised Interim Policy on Regulatory Structures for Low-Level Radioactive Waste Management and Disposal, and the DOE Implementation Plan for the Defense Nuclear Facility Safety Board (DNFSB) recommendation 94-2 define the requirements for management of LLW. This project provides LLW volume reduction, where possible, through incineration, compaction, and size reduction at the Waste Experimental Reduction Facility (WERF) and to disposal in the active pit of the Radioactive Waste Management Complex (RWMC) Subsurface Disposal Area (SDA). Other DOE Complex or commercial LLW disposal facilities will be utilized after FY2006 for LLW. This project will also plan and coordinate disposition of a small quantity of Special Case Waste (SCW). Waste generators will pay for the actual SCW disposition.

This project also supports RCRA treatment and disposal of HW using commercial TSD facilities, and energy recovery (cuber) of industrial waste to minimize volume of waste disposed.

Treatment, storage, and disposal of MLLW and LLW will decrease human and environmental risk by eliminating the waste stream backlog. Managing the waste in compliance with Federal, State, and DOE regulations reduces personnel exposure to these waste streams. Approved methods for treatment of the waste streams are used in preparation for disposal at approved waste depositories. Long term storage of waste containers will be minimized. The DNFSB Recommendation 94-2 Corrective Action Plan (INEL-96/0261A) addresses the ES&H vulnerabilities identified by the Complex wide review of LLW operations are corrected by this project.

This PBS is sufficiently funded to comply with Federal, State, and local regulations. Failure to comply with the regulatory drivers described above makes the INEEL liable for civil fines and penalties. This project will be followed by the Long Term Treatment/Storage/Disposal Operations project (ID-WM-107) and the AMWTP Production Operations project (ID-WM-105) summarized in Section A.2.6.

Project Baseline Summary Report

Report ID Number: Q501

Data Version: **16-Jan-98** Print Date: **19-Feb-98**

Operations/Field Office: **Idaho** Site: **Idaho National Engineering and Environmental Laboratory**

HQ ID: **IDIN0570** Project: **INEEL Low-Level Waste/Mixed Low-Level Waste/ Other Waste Program (ID-WM-101)**

A.2. Technical and Scope Narratives

A.2.2. Definition of Scope:

This project has one primary objective and five secondary objectives. The primary objective is to provide INEEL TSD services for MLLW until a demonstrated, more cost effective, commercial TSD is available to treat MLLW. Current plans call for the AMWTP to start treatment operations in March 2003. Capacity is being designed into the system to handle MLLW along with the transuranic waste. Upon successful demonstration of the AMWTP capability, WROC MLLW treatment activities will be suspended in September 2003.

Secondary objectives include: 1) Provide volume reduction and disposal of INEEL generated LLW through FY2006; 2) Establish off-site LLW disposal agreements/contracts at other DOE or commercial sites to support LLW disposal once the RWMC SDA active pit is filled; 3) Provide centralized planning and coordination for INEEL Special Case Waste (SCW) disposition; 4) Coordinate TSD services for INEEL generated HW; 5) Process INEEL combustible industrial waste into feed for the Idaho Chemical Processing Plant (ICPP) coal-fired steam generating plant.

The INEEL will focus on using the WERF incinerator to treat INEEL generated MLLW along with scheduling the excess capacity for other DOE sites MLLW. Ten incineration campaigns are planned each fiscal year at WERF. This approach is consistent with the DOE complex EM Integration Team, and in accordance with the INEEL STP. Compliance with the STP and RCRA will require:
- Operation of four MLLW treatment processes (incineration, stabilization, repackaging booth, and lead cask dismantlement/bulk lead treatment and disposal);
- Operation of four RCRA permitted storage facilities (PER-623 WERF Waste Storage Building (WWSB), PER-613 Mixed Waste Storage Facility (MWSF), Idaho Chemical Processing Plant (ICPP)-1617 Radioactive Mixed Waste Storage Facility, ICPP-1619 Hazard Chemical and Radioactive Waste Storage Facility);
- Maintain the INEEL emergency supply of bulk lead brick, sheet, and shot (PER-612 WROC Lead Storage Facility [WSLF]);
- Construction/operation of two new skid-mounted type treatment processes (macroencapsulation and sizing/sorting/segregation);
- Other DOE Complex or commercial treatment/disposal facilities will be used to support compliance with the STP. Examples include the DOE Oak Ridge Toxic Substance Control Act (TSCA) incinerator and the RCRA Subtitle C disposal facility operated by Envirocare in Utah.

The focus of the secondary objectives is to: 1) conduct LLW volume reduction through 2003 and disposal of INEEL generated LLW through 2006. Corrective actions identified in the DOE Implementation Plan for the DNFSB Recommendation 94-2 will be completed which will support continued environmentally safe LLW disposal

Project Baseline Summary Report

Report ID Number: Q501

Data Version: **16-Jan-98** Print Date: **19-Feb-98**

Operations/Field Office: **Idaho** Site: **Idaho National Engineering and Environmental Laboratory**

HQ ID: **IDIN0570** Project: **INEEL Low-Level Waste/Mixed Low-Level Waste/ Other Waste Program (ID-WM-101)**

A.2. Technical and Scope Narratives

through 2006. 2) Centralized planning to disposition SCW will be coordinated with the waste generators. SCW will require continued storage until disposal options are available. 3) Commercial TSD facilities will continue to be utilized for hazardous waste through FY2006. 4) Operation of the cuber will continue to process industrial waste through FY2006. The processing schedules for MLLW, LLW, HW, and industrial waste are described in Section A.4.

A.2.3. Technical Approach:

The overall approach for MLLW, LLW, SCW, HW, and industrial waste is to utilize the most cost effective option available. As commercial treatment and disposal capabilities become available and are proved cost effective, they will be used whenever possible, followed by existing INEEL or DOE Complex treatment units.

When treatment capability for specific MLLW streams is not available, new units will be designed and constructed (i.e., macroencapsulation). These new treatment processes will be designed for batch processing and have a small treatment capacity (tens of cubic meters per year). Small skid mounted treatment units will be constructed and placed into existing confinement areas for operation. Several treatment processes will be operated within the same confinement area within a given year. This represents a very low capital and cost effective approach to eliminating mixed waste streams at the INEEL in full compliance with the STP enforceable milestones.

A secondary advantage of MLLW treatment is LLW volume reduction. WERF incinerator operates continuously (24 hours per day/7 days per week) for approximately two weeks per month. During incineration of characteristic MLLW, the waste feed is supplemented with LLW in order to maintain incinerator operating temperatures. The resulting ash meets the criteria for disposal as LLW (either directly or following stabilization). This provides a dual benefit in that no surrogate material (e.g., clean feed stock such as corncobs, plastic, or oil used to increase the BTU content of the waste feed) must be purchased for supplemental waste feed and the LLW is treated for no additional cost. Listed MLLW is similarly augmented with LLW. The principal difference is that the amount of LLW is minimized because the resulting ash remains listed MLLW and requires offsite disposal at a Subtitle C facility.

Further LLW volume reduction is accomplished with the same operations staff required for MLLW incineration. When the incinerator is down for ash clean-out or maintenance, the same operational staff operates other MLLW treatment units or LLW size reduction and compaction processes. This provides significant LLW volume reduction, maximizing the effective use of the RWMC SDA active pit space for no additional labor costs.

A.2. Technical and Scope Narratives

Special Case Waste is generally not acceptable for near-surface disposal and has limited or no planned disposal alternative. SCW activities at the INEEL, within this PBS, are limited to a coordination effort for the SCW generators. Efforts include inventory of known SCW volumes, and coordinating generator treatment/temporary storage options.

Hazardous Waste (HW) will be consolidated in storage facilities or at the generating facility, awaiting treatment /disposal at an off site facility. Off site treatment/disposal facilities will be evaluated in support of direct shipment from the INEEL generator to the treatment/disposal vendor, thereby reducing the need for on site HW storage needs.

Industrial Waste cuber operations will continue in support of alternate fuel source for the coal fired steam generating facility at ICPP and reduced the volume disposed at the INEEL landfill.

Future technology development opportunities have been identified for advanced air pollution control methods including polishing capability for removal of dioxins, mercury, toxic metals, nitrogen and sulfur oxides, and hazardous hydrocarbons. Although none of these advanced technologies are required to support compliance with current State and Federal regulations and permits, they may result in increased throughput, reduced costs, or enhanced monitoring and will be pursued where practical. EPA`s new MACT Rule may require enhanced mercury and dioxin controls or monitoring at WERF. STCG Number 3.2.32, "Develop Thermal Treatment Unit Offgas CEM Monitors" and "Dioxin and Mercury Control for Incinerator Emissions for MACT Compliance" (STCG Number 3.1.31) are specific examples of these types of opportunities this PBS is pursuing.

A.2.4. Project Status in FY 2006:

The backlog of MLLW associated with this PBS will be treated and disposed by 2003. WROC MLLW and LLW treatment processes will be shut down in 2003. RCRA closure of WERF, the Repackaging Booth and two hazardous and MLLW storage facilities will be performed from 2004 through 2005 and is included in PBS ID-ER-110 - Decontamination & Decommissioning (D&D).

The backlog of contact handled LLW will be volume reduced and disposed by 2003. The Environmental Restoration (ER) and D&D programs will utilize the remaining capacity such that the active RWMC SDA disposal pit is predicted to be full by the year 2006 and will be ready for closure. RWMC SDA closure is included in PBS ID-ER-

Project Baseline Summary Report

Report ID Number: Q501

Data Version: **16-Jan-98** Print Date: **19-Feb-98**

Operations/Field Office: **Idaho** Site: **Idaho National Engineering and Environmental Laboratory**

HQ ID: **IDIN0570** Project: **INEEL Low-Level Waste/Mixed Low-Level Waste/ Other Waste Program (ID-WM-101)**

A.2. Technical and Scope Narratives

110 - Decontamination & Decommissioning (D&D). Volume reduction between 2004 and 2006 will use commercial facilities. The selected offsite disposal facility approved Waste Certification Programs and Waste Stream profiles will be in place by end of FY2004 for disposal of offsite contact handled (CH) LLW. CH waste will be actively disposed at the selected offsite disposal facility by FY2006. Preparations for RH waste disposal offsite will be in place. Issues will have been resolved regarding disposal offsite. The DOE Programmatic Impact Statement on Waste Management Activities will be issued and the path forward will be established. Cost/benefit studies will be completed. Continued onsite disposal of RH LLW may continue past FY2006. PBS #ID-WM-107 – Long Term Treatment/Storage/disposal Operations will perform this activity. After 2006, offsite disposal of LLW will be under PBS #ID-WM-107 – Long Term Treatment/Storage/disposal Operations.

The majority of SCW sealed sources will have been transferred to consolidated onsite storage and/or recycled offsite by FY2006. For other SCW, the generators will have completed characterization and the requirements for shipping and disposal will be identified and included in outyear funding requests.

HW and industrial waste will continue to be treated and disposed as it is generated. No backlog is anticipated.

A.2.5. Post 2006 Project Scope:

MLLW generation will continue for the life of the INEEL. Operation of the remaining MLLW storage facilities, along with treatment of newly generated MLLW by the AMWTP will be transferred to PBS ID-WM-107, Long Term Treatment/Storage/Disposal Operations beginning in 2007.

LLW generation will continue for the life of the INEEL. Commercial LLW volume reduction and offsite disposal of newly generated waste will be transferred to PBS ID-WM-107, Long Term Treatment/Storage/Disposal Operations beginning in 2007.

Centralized planning and coordination of SCW will be transferred to PBS ID-WM-107, Long Term Treatment/Storage/Disposal Operations beginning in 2007. The waste generators will be responsible for actual disposition costs.

HW generation will continue for the life of the INEEL. Commercial treatment and disposal facilities will continue to be utilized. Operation of the remaining hazardous waste storage facilities and shipment coordination services will be transferred to PBS ID-WM-107, Long Term Treatment/Storage/Disposal Operations beginning in 2007.

Project Baseline Summary Report

Report ID Number: Q501

Data Version: **16-Jan-98** Print Date: **19-Feb-98**

Operations/Field Office: **Idaho** Site: **Idaho National Engineering and Environmental Laboratory**

HQ ID: **IDIN0570** Project: **INEEL Low-Level Waste/Mixed Low-Level Waste/ Other Waste Program (ID-WM-101)**

A.2. Technical and Scope Narratives

Industrial waste generation will continue for the life of the INEEL. Operation of the ICPP cuber for industrial waste will be transferred to PBS ID-WM-107, Long Term Treatment/Storage/Disposal Operations beginning in 2007.

A.2.6. Project End State

MLLW, LLW, SCW, HW, and industrial waste generation will continue for the life of the INEEL. A significant portion of these wastes will be dispositioned within the 2006 Plan period; however, some services will extend up to FY2050. The final end state is to have all waste treated and disposed. Buildings will have been turned over to other programs for demolition or reuse. No legacy waste issues will remain.

Treatment of the MLLW backlog associated with this PBS was completed in 2003. Portions of the INEEL STP dealing with WROC MLLW treatments are marked complete. WROC MLLW treatment facilities and two hazardous waste and MLLW storage facilities were closed under RCRA (beginning in 2004). The remaining storage facilities were closed (beginning in 2011) when consolidated hazardous waste and MLLW storage was implemented within a Type II storage module at the RWMC. Buildings have been turned over for demolition or reuse. MLLW will be generated on the INEEL as long as nuclear operations continue. Current activities and future programs are expected to generate MLLW through 2050. Future generation of MLLW will be treated by the AMWTP.

The RWMC SDA CH LLW active disposal cell has been filled and the area was closed (beginning in 2007). LLW will be generated on the INEEL as long as nuclear operations continue. Current activities and future programs are expected to generate LLW through 2050. LLW volume reduction and disposal operations will be conducted at an offsite DOE or commercial facility. Special case waste has been dispositioned, primarily through shipment of material to an offsite geologic repository.

Hazardous waste will be generated in limited amouts due to the close of operations at the INEEL. Hazardous waste generated during D&D activities would be shipped directly from the generator to an off site treatment/disposal facility. HW storage facilities will be turned over for demolition or reuse.

Cuber operations are complete and the building has been turned over for demolition or reuse.

Project Baseline Summary Report

Report ID Number: Q501

Data Version: **16-Jan-98** Print Date: **19-Feb-98**

Operations/Field Office: **Idaho** Site: **Idaho National Engineering and Environmental Laboratory**

HQ ID: **IDIN0570** Project: **INEEL Low-Level Waste/Mixed Low-Level Waste/ Other Wastè Program (ID-WM-101)**

A.2. Technical and Scope Narratives

A.2.7. General Narrative:

Efforts are currently underway to evaluate closure of the RWMC SDA prior to 2006. Joint Waste Operations and Environmental Restoration task teams have been chartered to develop the strategy, along with a project based work plan to implement the strategy. The work plan (technical, cost, schedule) will then be integrated into specific 2006 Plan Project Baseline Summaries identified for the INEEL. The strategy will include optimization of the remaining capacity of the RWMC SDA based on: cost effectiveness, compliance with the PA limits, maintaining adequate capacity for critical customers, and filling the remaining capacity by 2003.

The LLW Quantity Table show disposition of the LLW backlog by the end of FY1999. This creates a significant spike in the quantity of sizable and non-volume reducable LLW requiring processing or shipment in FY1999. Current baseline funding does not support these values; however, efforts are underway to evaluate process changes which could result in increased throughput without significant increase in costs. Examples include: use of soft bags for disposal of large quantities of LLW and revision of the selection criteria for when it is cost effective for size reduction (i.e., do not size materials which give less than a 10 to 1 volume reduction).

A.2.8. Cost Baseline Narrative:

A detailed cost estimate was performed for each activity. The detailed estimates are for specific activities that must be performed to accomplish the project activities in full compliance with the Federal, State, and local regulations. The activities and costs were verified by a senior internal review board and rolled into a resource-loaded schedule that reflects current baseline compliance operations. Waste Operations is now in the process of projectizing activities to obtain further efficiencies. In completing the compliance baseline, an integral component of the projectization will to be to perform a critical analysis of our estimate by an independent review team. The cost estimates are based on FY1998 dollars with escalation of 2.7% applied annually on a compound basis to FY2006.

The cost baseline in this PBS does not include a charge back strategy for billing DOE sites for MLLW treatment services. This strategy may be modified once chargeback issues have been resolved throughout the complex.

A.2.9. Discuss How NEPA will or has been Address

Workscope described in this PBS is covered by the Department of Energy Programmatic Spent Fuel Management and Idaho National Engineering Laboratory Environmental Restoration and Waste Management Programs Final Environmental Impact Statement (DOE/EIS-0203-F) April 1995, and associated Record of Decision, May 1995. One

Data Version: **16-Jan-98** Print Date: **19-Feb-98**

Operations/Field Office: **Idaho** Site: **Idaho National Engineering and Environmental Laboratory**

HQ ID: **IDIN0570** Project: **INEEL Low-Level Waste/Mixed Low-Level Waste/ Other Waste Program (ID-WM-101)**

A.2. Technical and Scope Narratives

future action (offsite disposal of LLW at another DOE Complex or commercial facility) is dependent on decisions made in the Department of Energy Waste Management Programmatic Environmental Impact Statement. Individual projects are reviewed prior to implementation to ensure that adequate NEPA documentation exists or supplemental NEPA documentation is prepared.

A.2.10. 1997 Actual Accomplishments:

- The quantity of incinerable mixed waste treated in FY1997 was 52 m3 (original container volume). This waste had a repackaged volume of 286.5 m3 and weight of 87,221 lbs. There were six mixed waste burn campaigns, including two burn campaigns to treat off site waste. The RCRA Mini and Trial Burns were accomplished. Not all of the targets for the high temperature portion of the Trial Burn were achieved (destruction removal efficiency (DRE) for chlorobenzene was slightly low in one of the three runs). A second high temperature Trial Burn will be performed in FY1998.
- WROC completed the five scheduled INEL Site Treatment Plan milestones during FY1997. The MLLW Repackaging Booth Commence System Testing (P-4) and Commence Operations (P-5) milestones were both completed three weeks ahead of schedule. The INEEL Lead Program completed the Lead Cask Dismantlement Backlog P 6-1 and P-6-2 milestone 18 months ahead of schedule (61 m3). The incineration Backlog Schedule (P-6) milestone was completed on schedule.
- WROC supported the DOE, EM50 Cooperative Agreement and the DOE-ID/U.S.Army, Rock Island, Intra Agency Agreement and shipped approximately 39.4 m3 of contaminated lead to Envirocare of Utah for disposal.
- WROC completed 14 treatability studies in FY1997. Currently only 12 stored mixed waste streams remain that require treatablity studies.
- The 1997 MLLW first half Performance Measure Metrics Line C. New Waste includes an adjustment (74.5 m3) from what was previously submitted. This also increases Line A. Storage - Total Inventory.
- The second half Performance Measure Metrics Line C. New Waste includes 61 m3 of INEEL generated MLLW plus 15.1 m3 recieved from Los Alamos for incineration. Line D. Treatment includes: 1 m3 of incineration, 0.5 m3 from ICPP debris treatment, 17.9 m3 from the ANL-W Sodium Processing Facility, and 61 m3 of cask dismantlement (this volume will be reflected in Line G once the lead is recycled) . Line G. Volume Reduced is composed of two parts: 1) 57.2 m3 inventory reduction, 2) 19 m3 of contaminated lead sent to Manufacturing Sciences Corporation (MSC) for reuse in manufacturing shielded waste containers.
- WROC volume reduced 4,324 m3 of low level waste in FY1997, using the sizing, compaction and incineration process.
- Approximately 1,400 m3 of LLW was disposed of at the RWMC Sub Surface Disposal Area in FY1997.
- Low Level Waste Value Engineering Study (report issued 6/96, revised 12/96) issues were successfully closed out in FY1997. Long term items were transferred to other

Project Baseline Summary Report

Report ID Number: Q501

Data Version: **16-Jan-98** Print Date: **19-Feb-98**

Operations/Field Office: **Idaho** Site: **Idaho National Engineering and Environmental Laboratory**

HQ ID: **IDIN0570** Project: **INEEL Low-Level Waste/Mixed Low-Level Waste/ Other Waste Program (ID-WM-101)**

A.2. Technical and Scope Narratives

management tracking systems.

- Low Level Waste Vulnerability Assessment Corrective Action Plan deliverables for FY1997 were completed with the exception of the Composite Analysis. LMITCO and DOE-ID are working with the Naval Reactors Facility and Argonne National laboratories to obtain past disposal information that is to be included in the Composite Analysis. The SDA Performance Assessment was completed and submitted to DOE for review and approval.
- The Industrial Waste Program accomplished collection, monitoring, handling, shredding, cubing, and disposal of approximately 2,120 m3 of solid industrial waste.

A.2.11. 1998 Planned Accomplishments:

- Perform treatability studies on ten MLLW streams in support of MLLW treatment operations.
- Complete construction and initiate operations of the macroencapsulation (43 m3) and sizing/opening/segregation (60 m3) MLLW treatment units.
- Continue lead cask dismantlement activities at a reduced level from FY1997.
- Treat waste at WERF greater than 70 percent of the time (24 hrs per day, 365 days per year).
- Conduct LLW incineration (1416 m3), sizing (708 m3), and compaction (900 m3) operations on INEEL generated waste.
- Incinerate INEEL and other DOE sites MLLW (439 m3 original container volume).
- Stabilize WERF ash and other MLLW using Portland cement or other compatible materials 40 m3).
- Continue to expand MLLW/LLW production capability by increasing WERF operator certifications on MLLW/LLW treatment processes.
- Perform stabilization demonstration in conjunction with the Mixed Waste Focus Area using the ANL-E developed phosphate bonded ceramic process.
- Operate and maintain the MWSF, WWSB, CPP-1617, and CPP-1619 for hazardous waste and MLLW including interfacing with INEEL users.
- Maintain the INEEL lead emergency shielding reserve in the WLSF.
- Dispose up to 1800 m3 of LLW in the SDA.
- Perform activities to support the implementation of DNFSB recommendation 94-2.
- Submit the performance assessment / Composite Analysis report to DOE-HQ for the RWMC SDA.
- Submit annual report to DOE-HQ on Summary of Waste Disposal Operations and Performance Assessment Adequacy for the SDA.
- Conduct limited analyses, such as C-14 monitoring, H-3 monitoring and perched water and moisture migration monitoring to support PA dose calculations for the SDA.
- Provide overall strategic planning, technical waste evaluations and facilitate coordination between program and facility owners of SCW in continuing characterization,

A.2. Technical and Scope Narratives

preparation for interim storage and final geologic disposal.
- Operate the industrial waste cuber for onsite and Idaho Falls INEEL facilities generated wastes (5738 m3).

A.2.12. 1999 Planned Accomplishments:

- Continue lead cask dismantlement activities. Complete P6-3 Lead Cask dismantlement 75% treatment STP milestone.
- Continue MLLW lead treatment/disposal at Envirocare, Utah.
- Treat waste at WERF greater than 70 percent of the time (24 hrs per day, 365 days per year).
- Conduct LLW incineration (2748 m3), sizing (3032 m3), and compaction (400 m3) operations on INEEL generated waste.
- Incinerate INEEL and other DOE sites MLLW (439 m3 orginal container volume).
- Stabilize WERF ash and other MLLW using Portland cement or other compatible materials (40 m3).
- Operate the macroencapsulation (43 m3) and sizing/opening/segregation (60 m3) MLLW treatment units.
- Continue to expand MLLW/LLW production capability by increasing WERF operator certifications on MLLW/LLW treatment processes.
- Operate and maintain the MWSF, WWSB, CPP-1617, and CPP-1619 for hazardous waste and MLLW including interfacing with INEEL users.
- Maintain the INEEL lead emergency shielding reserve in the WLSF.
- Dispose up to 1800 m3 of LLW in the SDA;
- Perform PA validation studies based on Operations, D&D and Environmental Restoration activities;
- Activities to establish site specific release and transport rates for radionuclides will continue, and long-term waste generation projection rates will be updated.
- Provide overall strategic planning, technical waste evaluations and facilitate coordination between program and facility owners of SCW. Provide focused planning on Special Performance Assessment Required (SPAR) SCW characterization needs.
- Operate the industrial waste cuber for onsite and Idaho Falls INEEL facilities generated wastes (7650 m3).

Project Baseline Summary Report

Report ID Number: Q501

Data Version: **16-Jan-98** Print Date: **19-Feb-98**

Operations/Field Office: **Idaho** Site: **Idaho National Engineering and Environmental Laboratory**

HQ ID: **IDIN0570** Project: **INEEL Low-Level Waste/Mixed Low-Level Waste/ Other Waste Program (ID-WM-101)**

A.2. Technical and Scope Narratives

A.2.13. 2000 Planned Accomplishments:

- Continue lead cask dismantlement activities.
- Continue MLLW lead treatment/disposal at Envirocare, Utah.
- Treat waste at WERF greater than 70 percent of the time (24 hrs per day, 365 days per year).
- Conduct LLW incineration (1094 m3), sizing (307 m3), and compaction (413 m3) operations on INEEL generated waste.
- Incinerate INEEL and other DOE sites MLLW (439 m3 orginal container volume).
- Operate the macroencapsulation (43 m3) and sizing/opening/segregation (60 m3) MLLW treatment units.
- Stabilize WERF ash and other MLLW using Portland cement or other compatible materials (40 m3).
- Operate and maintain the MWSF, WWSB, CPP-1617, and CPP-1619 for hazardous waste and MLLW including interfacing with INEEL users.
- Maintain the INEEL lead emergency shielding reserve in the WLSF.
- Dispose up to 1800 m3 of LLW in the SDA.
- Provide overall strategic planning, technical waste evaluations and facilitate coordination between program and facility owners of SCW. Provide focused characterization and preparation planning of SPAR SCW for removal from storage pools to support D&D schedules for wet storage facilities.
- Operate the industrial waste cuber for onsite and Idaho Falls INEEL facilities generated wastes (7650 m3).
- Revise RWMC Performance Assessment and submitt to DOE-ID/DOE-HQ for approval.
- Collect data on corrsion tests at the RWMC SDA and issue report.
- Issue report on the column tests being conducted to identify subsurface transport rates.
- Issue radionuclide data report from the generator characterization improvement study for ICPP.

Project Baseline Summary Report

Report ID Number: Q501

Data Version: **16-Jan-98** Print Date: **19-Feb-98**

Operations/Field Office: **Idaho** Site: **Idaho National Engineering and Environmental Laboratory**

HQ ID: **IDIN0570** Project: **INEEL Low-Level Waste/Mixed Low-Level Waste/ Other Waste Program (ID-WM-101)**

A.2.15. Baseline Costs (in thousands of dollars)

Date Submitted: 12/12/97

	1997-2006 Total	2007-2070 Total	Grand Total	Planned 1997	Actual 1997	1998	1999	2000	2001	2002	2003	2004	2005	2006	2007
Current Cost Baseline	209,606	0	209,606	21,908	23,615	22,011	27,232	24,882	21,751	21,626	22,460	17,053	16,808	13,875	
Const 98 Baseline	193,490	0	193,490	22,500	24,253	22,011	26,516	23,591	20,080	19,440	19,659	14,534	13,948	11,212	
Storage	192,898	0	192,898	21,908		22,011	26,516	23,591	20,079	19,439	19,660	14,534	13,948	11,212	
Const 98 Storage	178,937	0	178,937	22,500		22,011	25,819	22,367	18,537	17,474	17,208	12,387	11,575	9,060	

	2008	2009	2010	2011-2015	2016-2020	2021-2025	2026-2030	2031-2035	2036-2040	2041-2045	2046-2050	2051-2055	2056-2060	2061-2065	2066-2070
Current Cost Baseline															
Const 98 Baseline															
Storage															
Const 98 Storage															

A.2.16. Non EM Costs included in the Cost Baseline

	1997	1998	1999	2000	2001	2002	2003	2004	2005	2006	2007	2008	2009
Environmental Management	100.00%	100.00%	100.00%	100.00%	100.00%	100.00%	100.00%	100.00%	100.00%	100.00%	100.00%	100.00%	100.00%

	2010	2011-2015	2016-2020	2021-2025	2026-2030	2031-2035	2036-2040	2041-2045	2046-2050	2051-2055	2056-2060	2061-2065	2066-2070
Environmental Management	100.00%	100.00%	100.00%	100.00%	100.00%	100.00%	100.00%	100.00%	100.00%	100.00%	100.00%	100.00%	100.00%

Project Baseline Summary Report

Report ID Number: Q501

Data Version: **16-Jan-98** Print Date: **19-Feb-98**

Operations/Field Office: **Idaho** Site: **Idaho National Engineering and Environmental Laboratory**

HQ ID: **IDIN0570** Project: **INEEL Low-Level Waste/Mixed Low-Level Waste/ Other Waste Program (ID-WM-101)**

A.2.17. Related Projects at the Same Site or Operations/Field Office

Project ID	Relation to this Project
ID-WM-105	Post FY2003 long term MLLW treatment
ID-WM-107	Post FY2006 LLW/MLLW management
ID-ER-110	Closure of facilities are performed in this PBS

A.2.18. Operations/Field Offices with Activities Related to this Project

Ops Office	Relation to this Project
Nevada	Off site LLW disposal at a DOE facility, transport of hazardous, radioactive, and classified material between projects.
Richland	Off site LLW disposal at a DOE facility, treatment of Hanford MLLW at WERF, transport of hazardous, radioactive, and classified material between projects.
Oak Ridge	K-25 incinerator for INEEL PCB waste, , treatment of Paducah MLLW at WERF
Albuquerque	Recycling specific Sealed Sources, transport of hazardous, radioactive, and classified material between projects, treatment of LANL, Sandia, and Pantex MLLW at WERF.
Headquarters	DOE Office of Civilian Radioactive Wastemanagement for SPAR SCW disposal with HLW
Chicago	Treatment of ANL MLLW at WERF, transport of hazardous, radioactive, and classified material between projects.
Oakland	Treatment of LBNL and LLNL MLLW at WERF, transport of hazardous, radioactive, and classified material between projects.
Rocky Flats	Treatment of Rocky Flats MLLW at WERF.

Project Baseline Summary Report

Report ID Number: Q501

Data Version: **16-Jan-98** Print Date: **19-Feb-98**

Operations/Field Office: **Idaho** Site: **Idaho National Engineering and Environmental Laboratory**

HQ ID: **IDIN0570** Project: **INEEL Low-Level Waste/Mixed Low-Level Waste/ Other Waste Program (ID-WM-101)**

A.2.18. Operations/Field Offices with Activities Related to this Project

Ops Office	Relation to this Project
Ohio	Treatment of Mound and West Valley MLLW at WERF.
	Treatment of Navy MLLW at WERF.
	Treatment of Weldon Springs MLLW at WERF.

A.2.19. Drivers:

CERCLA	RCRA	DNFSB	AEA	UMTRCA	State	DOE Orders	Other
Yes	Yes	Yes	Yes	No	Yes	Yes	No

A.2.20. Is this project A-106 (FEDPLAN) compliant? Yes

A.3. Milestones

Milestone/Activity	Field Milestone Code	Planned Date	Forecast Date	Actual Date	Status Indicator	EA	DNFSB	EM-1 or S-1	Intersite	HQ Change Control	Management Commitments	Key Decision
Project Start		10/1/96		10/1/96		No	No	No	No	No	No	No
Project Mission Complete		9/1/2006				No	No	No	No	No	No	No
LT S&M Completion (If applicable)		9/1/2050				No	No	No	No	No	No	No
P6 - Incineration - Backlog Schedule		3/1/97		3/1/97		Yes	No	No	No	No	No	No
P4 - Repackaging Booth - Commence System Testing		3/1/97		3/1/97		Yes	No	No	No	No	No	No

Project Baseline Summary Report

Report ID Number: Q501

Data Version: **16-Jan-98** Print Date: **19-Feb-98**

Operations/Field Office: **Idaho** Site: **Idaho National Engineering and Environmental Laboratory**

HQ ID: **IDIN0570** Project: **INEEL Low-Level Waste/Mixed Low-Level Waste/ Other Waste Program (ID-WM-101)**

A.3. Milestones

Milestone/Activity	Field Milestone Code	Planned Date	Forecast Date	Actual Date	Status Indicator	EA	DNFSB	EM-1 or S-1	Intersite	HQ Change Control	Management Commitments	Key Decision
Complete WERF RCRA Trial Burn		7/1/97		7/1/97		No	No	No	No	No	No	No
P5 - Repackaging Booth - Commence Operations		6/1/97		6/1/97		Yes	No	No	No	No	No	No
P3 - Macroencapsulation - Initiate Construction	234M104015	3/1/98				Yes	No	No	No	No	No	No
P3- Sizing/Opening/Segregation - Initiate Construction	234M104050	6/1/98				Yes	No	No	No	No	No	No
P4 - Macroencapsulation - Commence System Testing	234M101105	9/1/98				Yes	No	No	No	No	No	No
P2 - HG Retort - Procure Contracts		12/1/98			C	Yes	No	No	No	No	No	No
P6 - Repackaging Booth - Backlog Schedule	2320302205	3/1/98			C	Yes	No	No	No	No	No	No
P6 - Stabilization - Backlog Schedule	2320302250	3/1/98				Yes	No	No	No	No	No	No
P4 - Sizing/Opening/Segregation - Commence System Testing	234M101130	12/1/98				Yes	No	No	No	No	No	No
P6-1 - Cask Dismantlement - 25% Backlog Complete	23201011292	3/1/98		9/1/97		Yes	No	No	No	No	No	No
P1 - HG Retort - Submit RCRA Permit		6/1/98			C	Yes	No	No	No	No	No	No
P5 - Macroencapsulation -	2320201805	3/1/99				Yes	No	No	No	No	No	No

Project Baseline Summary Report

Report ID Number: Q501

Data Version: **16-Jan-98** Print Date: **19-Feb-98**

Operations/Field Office: **Idaho** Site: **Idaho National Engineering and Environmental Laboratory**

HQ ID: **IDIN0570** Project: **INEEL Low-Level Waste/Mixed Low-Level Waste/ Other Waste Program (ID-WM-101)**

A.3. Milestones

Milestone/Activity	Field Milestone Code	Planned Date	Forecast Date	Actual Date	Status Indicator	EA	DNFSB	EM-1 or S-1	Intersite	HQ Change Control	Management Commitments	Key Decision
Commence Operations												
P5 - Sizing/Opening/Segregation - Commence Operations	2320201810	6/1/99				Yes	No	No	No	No	No	No
P6 - Macroencapsulation - Backlog Schedule	2320302280	9/1/99				Yes	No	No	No	No	No	No
P3 -HG Retort - Initiate Construction		12/1/98			C	Yes	No	No	No	No	No	No
P6-2 Cask Dismantlement - 50% Backlog Complete	23201011294	3/1/99		9/1/97		Yes	No	No	No	No	No	No
P4 - HG Retort - Establish Contracts		12/1/2000				Yes	No	No	No	No	No	No
P6-3 - Cask Dismantlement - 75% Backlog Complete	23201011296	9/1/99				Yes	No	No	No	No	No	No
P6 - Sizing/Opening/Segregation - Backlog Schedule	2320302310	6/1/2000			C	Yes	No	No	No	No	No	No
P5 - HG Retort - Ship Waste Offsite for Treatment		3/1/2000				Yes	No	No	No	No	No	No
P6 - HG Retort - Submit Backlog Schedule for Offsite Treatment		6/1/2000				Yes	No	No	No	No	No	No

Project Baseline Summary Report

Report ID Number: Q501

Data Version: **16-Jan-98** Print Date: **19-Feb-98**

Operations/Field Office: **Idaho** Site: **Idaho National Engineering and Environmental Laboratory**

HQ ID: **IDIN0570** Project: **INEEL Low-Level Waste/Mixed Low-Level Waste/ Other Waste Program (ID-WM-101)**

A.4. Performance Measure Metrics

Category/Subcategory	Units	1997 Year End	
		Planned	Actual
III. MLLW			
A. Inv. - Total	M3	1,131.00	1,294.93
C. New Waste	M3	72.35	278.70
D. Treatment	M3	150.00	132.48
E. Disp. - On-site/Comm.	M3	106.50	53.26
F. Disp. - DOE Offsite	M3	0.00	3.24
G. Volume Reduced	M3	27.85	120.27
IV. LLW			
A. Inv. - Total	M3	12,843.03	9,731.27
C. New Waste	M3	3,627.41	2,623.41
D. Treatment	M3	3,977.34	4,298.61
E. Disp. - On-site/Comm.	M3	1,813.04	1,293.59
F. Disp. - DOE Offsite	M3	0.00	0.00
G. Volume Reduced	M3	1,763.00	8,367.55
V. HAZ			
A. Inv. - Total	MT	4.50	5.38
C. New Waste	MT	30.00	20.43
D. Treatment	MT	0.00	0.00
E. Disp. - On-site/Comm.	MT	40.00	29.55

Project Baseline Summary Report

Report ID Number: Q501

Data Version: **16-Jan-98** Print Date: **19-Feb-98**

Operations/Field Office: **Idaho** Site: **Idaho National Engineering and Environmental Laboratory**

HQ ID: **IDIN0570** Project: **INEEL Low-Level Waste/Mixed Low-Level Waste/ Other Waste Program (ID-WM-101)**

A.4. Performance Measure Metrics

Category/Subcategory	Units	1997 Year End	
		Planned	Actual
V. HAZ			
G. Volume Reduced	MT	0.00	0.00
VI. SAN			
C. New Waste	M3	4,013.92	5,970.41
E. Disp. - On-site/Comm.	M3	4,013.92	5,970.41

A.6. Validation

A.6.1. Project Validated? No **A.6.2. Date Validated:**

A.6.3. Validation Method:

A joint senior level DOE-ID and LMITCO Independent Murder Board Review of the INEEL decision units was conducted. Six teams consisting of six members reviewed the scope, schedule, cost estimates, and basis of estimates for each of the decision units which are the same base elements used to construct the PBS.

A.6.4. Technical Approach Reference Documents:

HAZARDOUS WASTE DRIVERS -
DOE: DOE 0231.1; DOE 435.1; DOE Suppl. Order 5400.1 Section 5.
EPA: RCRA Part B Permit; RRWAC DOE/ID 10831; INEEL Site Treatment Plan; National Environmental Policy Act (NEPA).
STATE: Idaho Hazardous Waste Management Act, Chap 44; Environmental Monitoring and Site Agreement; State Water Rights Agreement; Settlement Agreement.
CFR: 40 CFR 260.10; 40 CFR 261.3; 40 CFR 262; 40 CFR 264(I); 40 CFR 264.17; 40 CFR 268.7; 49 CFR 25; 49 CFR 173.24(e); 49 CFR 177.848.
LIQUID TOXICS (on-site discharges) DRIVERS -

Project Baseline Summary Report

Report ID Number: Q501

Data Version: **16-Jan-98** Print Date: **19-Feb-98**

Operations/Field Office: **Idaho** Site: **Idaho National Engineering and Environmental Laboratory**

HQ ID: **IDIN0570** Project: **INEEL Low-Level Waste/Mixed Low-Level Waste/ Other Waste Program (ID-WM-101)**

A.6. Validation

DOE: DOE 0231.1; DOE 5400.1; RRWAC DOE/ID 10831.
STATE: Idaho Water Quality Standards Wastewater Treatment Req`s Manual.
CFR: 40 CFR.
SOLID LOW-LEVEL WASTE DRIVERS -
DOE: DOE 0231.1; DOE 0460.1; DOE 435.1; Atomic Energy Act of 1954; DOE 5480.3; INEEL Radiological Control Manual DOE-ID 10399, Art. 422; RRWAC DOE/ID 10831.
EPA: Regulation LLWPAA 1985, PL 99-2; Federal Facilities Compliance Act (FFCA).
STATE: Settlement Agreement; INEEL Site Treatment Plan; WERF SAR; RWMC EDFs 484 & 485; RWMC SAR.
CFR: 10 CFR 20.2005; 10 CFR 61; 49 CFR 172.403; 49 CFR 173.
SOLID TRANSURANIC WASTE DRIVERS -
DOE: Atomic Energy Act of 1954; DOE 435.1.
STATE: Settlement Agreement; INEEL Site Treatment Plan.
MIXED WASTE DRIVERS -
DOE: DOE 0231.1; DOE 0460.1; DOE 435.1; DOE 5480.3; INEEL Radiological Control Manual DOE-ID 10399, Art 422; RRWAC DOE/ID 10831;
EPA: RCRA Part B Permit, Federal Facilities Compliance Act (FFCA) (180-day report); NEPA.
CFR: 40 CFR 262; 40 CFR 264; 40 CFR 268; 49 CFR 172/173.
STATE: Idaho Hazardous Waste Management Act Chapter 44; Environmental Monitoring Site Agreement; State Water Rights Agreement; Settlement Agreement; INEEL Site Treatment Plan.
TRANSPORTATION DRIVERS -
CFR: 49CFR100-180, 10CFR71, 10CFR 830.120, DOE Orders 5700.6C, 460.1A, 5632.11

A.6.5. Current Status of Project Baseline:

This project baseline currently reflects scope against the FY1998 Congressional appropriation.

A.6.6. Is the PBS Consistent with the Site Baseline? Yes

A.6. Validation

A.6.7. If PBS is Not Consistent with the Site Baseline, Why Not?

N/A

A.6.8. Future Validation Plans and Schedule:

Project Baseline will be revalidated during final review of PBS by DOE-ID prior to submittal to HQ. No further validation of Project Baseline is anticipated.

A.6.9. Site Baseline Consistency: 75% - PBS Well Supported by Site Baseline(s)

A.6.10. Project End State Definition: 75% - Project End State is Well Defined

A.7. Project Assumptions

Number	Assumption
1.	Oak Ridge, TN TSCA incinerator is available to treat INEEL TSCA waste at no cost to INEEL.
2.	The State of Idaho will approve RCRA Part B permit applications within one year of submittal.
3.	Advanced Mixed Waste Treatment Project (AMWTP) is operational in March 2003.
4.	Industrial waste collections and disposal are paid out of LMITCO indirect accounts.
5.	The cost baseline does not include facility modifications to support the proposed EPA Maximum Achievable Control Technology (MACT) rule. Significant modifications may be required.
6.	Left intentionally blank.
7.	A waste storage module at the RWMC is available for MLLW and HW storage in FY2010.

Project Baseline Summary Report

Report ID Number: Q501

Data Version: **16-Jan-98** Print Date: **19-Feb-98**

Operations/Field Office: **Idaho** Site: **Idaho National Engineering and Environmental Laboratory**

HQ ID: **IDIN0570** Project: **INEEL Low-Level Waste/Mixed Low-Level Waste/ Other Waste Program (ID-WM-101)**

A.7. Project Assumptions

Number	Assumption
8.	Offsite MLLW generators will not be charged for treatment at WERF. Implementation of a waste generator chargeback process will be re-evaluated once the requirements are better defined across the DOE Complex.
9.	Left intentionally blank.
10.	The composite analysis required by the Defense Nuclear Facility Safety Board will continue to allow for LLW disposal in the SDA until FY2006.
11.	Designation of INEEL as an offsite generator to an offsite DOE facility will be obtained from DOE-HQ prior to FY2004 regardless of the filled status of the active pits at the RWMC.
12.	Facility modifications required for RH waste, if required by Naval Reactor Facilities and ATR, will be supported by DOE-HQ as required to meet FY2006 RWMC shutdown schedule, and are not covered in this PBS.
13.	Non-EM generators will be financially responsible for cost of disposal of wastes at offsite facilities including waste packaging, characterization, transportation, and disposal.
14.	The current moratorium on offsite generators sending waste to the Nevada Test Site will be lifted, or a suitable alternative offsite disposal facility will be identified.
15.	SCW in the non-certifiable TRU and non-defense TRU waste subcategories are addressed (and costed) separately under the TRU Project.
16.	SCW in the fuel and fuel debris subcategory is addressed (and costed) separately under the Spent Nuclear Fuel Project.
17.	Co-disposal of SPAR SCW in a deep geological repository will be possible if proven to be cost effective and safe (no separate repository will be built for SCW).

Project Baseline Summary Report

Report ID Number: Q501

Data Version: **16-Jan-98** Print Date: **19-Feb-98**

Operations/Field Office: **Idaho** Site: **Idaho National Engineering and Environmental Laboratory**

HQ ID: **IDIN0570** Project: **INEEL Low-Level Waste/Mixed Low-Level Waste/ Other Waste Program (ID-WM-101)**

B.1. Budget by Appropriations Account (in thousands of current year dollars)

Appropriations Account	FY 1997 BA	FY 1998 BA	FY 1999 BA	FY 2000 BA
EM Defense	21,027	22,011	27,232	24,882

C.1. Risk

C.1.1. Risk Data:

Project risk is a concatenation of a number code and a letter code. The number code represents the level of impact with 1 being the greatest impact and 4 being the least impact. The letter code represents the likelihood of an event occurring (either probability of event or time until event) with A being the most likely and D being the least likely to occur. The risk code is followed by a U, H, M, L, or N (Urgent - N/A) representing the risk level of the project. For a more detailed description of the risk data see Section C.1.1. of the October 1998, 2006 Plan Guidance.

	1997	1998	1999	2000	2001	2002	2003	2004	2005	2006	2007	2008	2009	2010
Environment	3D - L	3D - L	3D - L	3D - L	3D - L	3D - L	3D - L	3D - L	3D - L	3D - L				
Public	3D - L	3D - L	3D - L	3D - L	3D - L	3D - L	3D - L	3D - L	3D - L	3D - L				
Worker	3D - L	3D - L	3D - L	3D - L	3D - L	3D - L	3D - L	3D - L	3D - L	3D - L				

	2011-2015	2016-2020	2021-2025	2026-2030	2031-2035	2036-2040	2041-2045	2046-2050	2051-2055	2056-2060	2061-2065	2066-2070
Environment												
Public												
Worker												

C.1.2. Choose either the public, worker, or the environment as the End-State Risk driver: Environment

C.1.3. Choose either the public, worker, or the environment as the Interim Risk driver: Worker

Project Baseline Summary Report

Report ID Number: Q501

Data Version: **16-Jan-98** Print Date: **19-Feb-98**

Operations/Field Office: **Idaho** Site: **Idaho National Engineering and Environmental Laboratory**

HQ ID: **IDIN0570** Project: **INEEL Low-Level Waste/Mixed Low-Level Waste/ Other Waste Program (ID-WM-101)**

C.1.4. If upon completion of this project, another project manages its hazards, indicate that project:	ID-WM-105 / AMWTP Production Operations
C.1.5. Has the risk evaluation been internally peer reviewed by ES&H professionals?	Yes
C.1.6. Has the risk evaluation been externally peer reviewed?	Yes
C.1.7. Have regulators, stakeholders, & Tribal Nations been involved in validating the project risk evaluations?	Yes

D.1. - Direct Safety & Health Narratives and Risk Narratives

D.1.2. Direct S &H Narrative - Hazards:

This project, currently in the operational phase, contains the S&H functions necessary to treat, store, and or dispose of mixed low level (MLLW), hazardous, low level (LLW) and industrial wastes at the INEEL. This project manages and operates the facilities necessary to perform the INEEL missions for the aforementioned waste streams. The facilities are located at three (RWMC, ICPP, and WROC/PBF) different areas on the INEEL at distances up to 10 miles. Hazards associated with the operation activities of this project include occupational, chemical, radiation exposure and risk to workers who operate in industrial facilities. Radiological and chemical exposure can occur during waste processing and material handling as well as a result of a fire or spill.

Hazards are documented and addressed in hazard analyses, Safety Analysis Reports, a Health and Safety Program, and operation documentation (i.e. Radiological Work Permit, Safe Work Permit, Confined Space Permit, etc.).

Hazards are mitigated by job planning and during operations by incorporating engineered controls (e.g. ventilation), the use of personnel protective equipment, monitoring, training, work procedures, and the INEEL ALARA Program. INEEL personnel have participated in the Voluntary Protection Program and are aware that their personnel safety begins with their own attitude.

At the end state of this project, the hazards are mitigated due to the completion of treatment and the disposal of the waste.

This project`s cask dismantlement activities pose an additional risk to personnel concerning the uptake of airborne lead particles. Although this PBS provides for dismantlement activities it does not eveluate the hazards that are associated with the activity. The INEEL Test Area North (TAN) performs these activities in accordance with LMITCO procedure MCP 2720 and has evaluated the industrial, radiological, and medical hazards associated with lead handling. The lead hazards, if above the Personnel Exposure Limit (PEL), will be documented in an internal Compliance Plan and are reviewed by first line management and cognizant S&H personnel.

Project Baseline Summary Report

Report ID Number: Q501

Data Version: **16-Jan-98**

Print Date: **19-Feb-98**

Operations/Field Office: **Idaho**

Site: **Idaho National Engineering and Environmental Laboratory**

HQ ID: **IDIN0570** Project: **INEEL Low-Level Waste/Mixed Low-Level Waste/ Other Waste Program (ID-WM-101)**

D.1. - Direct Safety & Health Narratives and Risk Narratives

D.1.3. Direct S&H Narrative - Controls:

Continued treatment/disposal of LLW/MLLW at the INEEL reduces the environmental and health risks as outlined in the Programmatic Spent Nuclear Fuel Management and INEL Environmental Restoration and Waste Management Programs Final Environmental Impact Statement (INEL EIS), Volume 2, table 5.14-3 (Alternate A) and Table 5.14-8 (Alternative D). In addition, facilities have current authorization bases, within this project, including facility specific Safety Analysis Reports, RCRA Interim Status/Permit status, the INEL Site Treatment Plan, Performance Assessment, and/or Hazard Analyses.

Personnel hazards are mitigated during day to day operations by implementation of a Conduct of Operations approach including elements like work control procedures, Compliance Plan, radiological hazard analysis, safe work permits and industrial safety and industrial hygiene monitoring. Proper use of personnel protective equipment is ensured through training, health and safety professional reviews, and job supervisor spot checks. The INEEL ALARA Program is fully implemented at the facility and personnel exposure is minimized. The medical bioassay program for monitoring chemical and radiological uptakes is provided within this project.

This project's hazard bases can be found in the following documents:

Hazard Analysis for Waste Experimental Reduction Facility; EGG-WM-11467, September 1994 and can be found in WROC Document Control, Building PER 601, Hazard Analysis for WROC mixed waste storage (MWSF/PSU); EGG-WM-11153, February 1994 and can be found in WROC Document Control, Building PER 601, Waste Experimental Reduction Facility Safety Analysis Report; INEL-96/0165 (WERF), August 1996 and can be found in WROC Document Control, Building PER 601, Mixed Waste Storage Facility Safety Analysis Report; EGG-WM-10896 (MWSF/PSU), Rev.1, January 1996 and can be found in WROC Document Control, Building PER 601, Idaho Chemical Processing Plant Safety Analysis Report; WIN-107-8-9, June 1994, and document can be found in ICPP Document Control, building CPP665, Radioactve Waste Management Complex Low Level Waste Radiological Performance Assessment, EGG-WM-8773, May 1994 .

Radioactve Waste Management Complex Safety Analysis Report, INEL-94/0226, Rev. 2, 7/1997.

D.1.4. Direct S&H Narrative - Work Performance:

The resources necessary to accomplish MLLW treatment and LLW volume reduction safely is provided through the funding authorization for this project. Resources necessary for S&H oversight for disposal of LLW at the Radioactive Waste Management Complex (RWMC) is supported by the tenant facility and PBS WM-103 - INEEL TRU Waste Program. S&H resources within this project are planned and resource loaded into the project management software on a life cycle bases.

Activities within this project have been classified as less than Category III under DOE Order 5480.22, therefore, new MLLW treatment activities do not require an Operational Readiness Assessment. The project will perform a Management Assessment of all new waste treatment processes prior to operations.

Project Baseline Summary Report

Report ID Number: Q501

Data Version: **16-Jan-98** Print Date: **19-Feb-98**

Operations/Field Office: **Idaho** Site: **Idaho National Engineering and Environmental Laboratory**

HQ ID: **IDIN0570** Project: **INEEL Low-Level Waste/Mixed Low-Level Waste/ Other Waste Program (ID-WM-101)**

D.1. - Direct Safety & Health Narratives and Risk Narratives

S&H resources necessary to accomplish MLLW treatment activities include pre job safety, radiological, and quality reviews; at the job S&H inspections; daily, weekly, and monthly surveys in S&H areas; continual hazard analysis of high personnel risk activities (ash handling, MLLW repackaging). There is no appreciable change in S&H resource requirements during the operational phase of this project. Upon completion of MLLW treatment that is schedule for this project, closure will commence.
Cask dismantlement at TAN and other lead handling activities are addressed by a LMITCO internal Compliance Plan . Industrial, radiological, and medical hazards are outlined by this document, as well as the protective equipment required. Contious monitoring of activities by work supervisors and cognizant professionals help mitigate the possibility of worker exposure to lead.
S&H resources necessary for RCRA closure are included in PBS ID-ER-110 - Decontamination & Decommissioning (D&D). S&H resources necessary for futue treatment of MLLW and disposal of LLW, after 2006, are included in the Long Term Treatment/Storage/Disposal Operations project (ID-WM-107) and the AMWTP Production Operations project (ID-WM-105).
The average cost per FTE assumed (burdened rate) is $85K /year for Industrial Safety, $82K/year for Industrial Hygiene, $89K/year Radiological Engineering, $65K/year for Radiological Control Technician, and $84K/year for Fire Protection.

D.1.5. Direct S&H Narrative - Feedback and Continuous Improvement:

S&H compliance is verified by continuous surveillance, tracking of deficiencies in the INEEL ICARE system, and operates an Administrative preventive Maintenance system to control facility status. Maintenance of the compensatory measures will also verify compliance. ES&H oversight assessments will be conducted annually and are provided for in this PBS. Implementation of the INEEL Voluntary Protection Program enables each employee to report and receive closure on items of concern they raise.

D.1.6. Risk Evaluation Narrative:

Continued treatment of hazardous and radiological contaminated wastes at the INEEL reduces the environmental and health risk as outlined in the Programmatic Spent Nuclear Fuel Management and INEL Environmental Restoration and Waste Management Programs Final Environmental Impact Statement (INEL EIS), Volume 2, Table 5.14-3 (Alternative A) and Table 5.14-8 (Alternative). Failure to mitigate existing risks would strongly impact public trust and confidence.

MLLW worst case scenario for treatment operations:
Fire in the WERF Waste Storage Building (WWSB); INEL EIS - Alternative D. The Maximum Treatment Alternative assumes treatment of INEL LLW and MLLW, along

D.1. - Direct Safety & Health Narratives and Risk Narratives

with some waste from other DOE Complex of Federal government installations.

Public Safety and Health:
Probability 1E-02/yr; MEI Dose 2.8E-03 rem; MEI Cancer Risk 1.4E-08 fatal cancers/yr.

Site Personnel Safety and Health:
In accordance with DOE-STD-1027, the Category 3 threshold limits for facility inventory of radionuclides is based on Reportable Quanities. DOE-STD-1027 specifies the sum of the ratios limit of one, the maximum dose possible at 30 meters from total facility inventory would be 500 mrem.

Environmental Impact:
Minor onsite environmental impacts due to above release of radioactive and hazardous materials. No offsite health effects.

Per the RWMC SAR, solid LLW disposed of at the SDA is in permanent burial/storage; waste disposed of at the SDA is not intended to be retrieved. The impacts of permanent LLW disposal on the environment and public are not evaluated in the RWMC SAR. LLW operation activity has a less probability and risk to the public than MLLW activities.

Project Baseline Summary Report

Report ID Number: Q501

Data Version: **16-Jan-98** Print Date: **19-Feb-98**

Operations/Field Office: **Idaho** Site: **Idaho National Engineering and Environmental Laboratory**

HQ ID: **IDIN0570** Project: **INEEL Low-Level Waste/Mixed Low-Level Waste/ Other Waste Program (ID-WM-101)**

D.2. - Safety and HealthDirect Data

D.2.2. Safety and Health Cost Reporting - Direct Costs (in thousands of current year dollars)

	1997	1998	1999	2000
A. Emergency Preparedness	86	77	70	70
B. Fire Protection	45	40	40	40
C. Industrial Hygiene	154	138	138	138
D. Industrial Safety	48	43	50	38
E. Occupational Medicine	0	0	0	0
F. Nuclear Safety	191	171	176	176
G. Radiation Protection	1,064	953	930	953
H. Transportation Safety	0	0	0	0
I. Management Oversight	305	273	273	273
Total S&H Direct Costs:	**1,893**	**1,695**	**1,677**	**1,688**

D.2.5. Safety and Health FTE Reporting - Direct Contractor FTEs

	1997	1998	1999	2000
A. Emergency Preparedness	0.74	0.73	0.66	0.66
B. Fire Protection	0.56	0.50	0.50	0.50
C. Industrial Hygiene	2.24	2.01	2.01	2.01
D. Industrial Safety	0.74	0.57	0.66	0.51
E. Occupational Medicine	0.00	0.00	0.00	0.00
F. Nuclear Safety	1.71	1.49	1.53	1.53
G. Radiation Protection	11.12	10.29	9.96	10.27
H. Transportation Safety	0.00	0.00	0.00	0.00
I. Management Oversight	2.28	2.04	2.04	2.04
Total Direct Contractor FTEs:	**19.39**	**17.63**	**17.36**	**17.52**

Project Baseline Summary Report

Report ID Number: Q501

Data Version: **16-Jan-98** Print Date: **19-Feb-98**

Operations/Field Office: **Idaho** Site: **Idaho National Engineering and Environmental Laboratory**

HQ ID: **IDIN0570** Project: **INEEL Low-Level Waste/Mixed Low-Level Waste/ Other Waste Program (ID-WM-101)**

E. Enhanced Performance Measures

E.1. Project Estimates (thousands of current year dollars)

E.1.1. Current Estimated Lifecycle Cost of Project:	209,606	**E.1.2. Previously Estimated Lifecycle Cost of Project:**	
E.1.3. Projected Cost for FY 97:	21,908	**E.1.4. Projected % Work Completed by End of FY98:**	
E.1.5. Current Projected End Date of Project:	01-Sep-06	**E.1.6. Previously Projected End Date of Project:**	

E.2. Performance for FY 1997 (thousands of current year dollars

E.2.1. Actual Cost for FY 97:	23,615	**E.2.2. Actual % Work Completed to date:**	

Appendix C

List of Geographic Sites

Appendix C. List of Geographic Sites

The following tables list 134 geographic sites (including the Waste Isolation Pilot Plant) that EM has historically included in its scope. Following are four tables:

1. Completed prior to 1997 (Table C.1)

   23 FUSRAP[1] sites

   16 UMTRA[2] sites (long-term surveillance and monitoring and groundwater monitoring included in *Paths to Closure*)

   11 Other sites (long-term surveillance and monitoring as required included in *Paths to Closure*)

   50 TOTAL SITES COMPLETED PRIOR TO 1997

2. Completed in 1997 (Table C.2)

   2 FUSRAP sites

   4 UMTRA sites (included in *Paths to Closure*)

   4 Other sites (included in *Paths to Closure*)

   10 TOTAL SITES COMPLETED IN 1997

3. Transferred to the United States Army Corps of Engineers (Table C.3)

   21 FUSRAP Sites

   21 TOTAL SITES TRANSFERRED TO THE UNITED STATES ARMY CORPS OF ENGINEERS

4. Sites remaining (all covered in *Paths to Closure*) (Table C.4)

   0 FUSRAP sites

   4 UMTRA sites (including Belfield and Bowman, which were delisted in 1998)

   49 other sites

   53 TOTAL SITES REMAINING

Paths to Closure addresses all completed EM sites for which EM is responsible for long-term surveillance and monitoring from Table C.1.

Paths to Closure also addresses all sites that still required cleanup as of the beginning of FY 1997 (except for the two FUSRAP sites completed in FY 1997).

[1]Formerly Utilized Sites Remedial Action Program
[2]Uranium Mill Tailings Remedial Action

Table C.1
Sites Completed Prior to 1997

State	Operations/ Field Office	Site	Completion Date
Alaska	Nevada	Project Chariot (Nevada Offsite)	completed
Arizona	Albuquerque	Monument Valley (UMTRA site)	completed
Arizona	Albuquerque	Tuba City (UMTRA site)	completed
California	Albuquerque	Oxnard Facility	completed
California	Albuquerque	Salton Sea Test Base	completed
California	Oak Ridge	University of California (FUSRAP site)	completed
Colorado	Albuquerque	Durango (UMTRA site)	completed
Colorado	Albuquerque	Grand Junction Mill Tailings Site (UMTRA site)	completed
Colorado	Albuquerque	Gunnison (UMTRA site)	completed
Connecticut	Oak Ridge	Seymour Specialty Wire (FUSRAP site)	completed
Florida	Albuquerque	Peak Oil PRP Participation	completed
Hawaii	Albuquerque	Kauai Test Facility	completed
Idaho	Albuquerque	Lowman (UMTRA site)	completed
Illinois	Oak Ridge	Granite City Steel (FUSRAP site)	completed
Illinois	Oak Ridge	National Guard Armory (FUSRAP site)	completed
Illinois	Oak Ridge	University of Chicago (FUSRAP site)	completed
Massachusetts	Oak Ridge	Chapman Valve (FUSRAP site)	completed
Michigan	Oak Ridge	General Motors (FUSRAP site)	completed
Nebraska	Chicago	Hallam Nuclear Power Facility	completed
New Jersey	Oak Ridge	Kellex/Pierpont (FUSRAP)	completed
New Jersey	Oak Ridge	Middlesex Municipal Landfill (FUSRAP site)	completed
New Mexico	Oak Ridge	Acid/Pueblo Canyons (FUSRAP site)	completed
New Mexico	Albuquerque	Ambrosia Lake (UMTRA site)	completed
New Mexico	Oak Ridge	Bayo Canyon (FUSRAP site)	completed
New Mexico	Oak Ridge	Chupadera Mesa (FUSRAP site)	completed
New Mexico	Albuquerque	Holloman AFB	completed

Table C.1
Sites Completed Prior to 1997 (Continued)

State	Operations/ Field Office	Site	Completion Date
New Mexico	Albuquerque	Pagano Salvage Yard	completed
New Mexico	Albuquerque	Shiprock (UMTRA site)	completed
New Mexico	Albuquerque	South Valley Superfund Site	completed
New York	Oak Ridge	Baker and Williams Warehouses (FUSRAP site)	completed
New York	Oak Ridge	Niagara Falls Storage Site Vicinity Properties (FUSRAP site)	completed
Ohio	Oak Ridge	Alba Craft (FUSRAP site)	completed
Ohio	Oak Ridge	Associate Aircraft (FUSRAP site)	completed
Ohio	Oak Ridge	B&T Metals (FUSRAP site)	completed
Ohio	Oak Ridge	Baker Brothers (FUSRAP site)	completed
Ohio	Oak Ridge	Herring-Hall Marvin Safe Company (FUSRAP site)	completed
Ohio	Chicago	Piqua, Ohio Site	completed
Oregon	Oak Ridge	Albany Research Center (FUSRAP site)	completed
Oregon	Albuquerque	Lakeview (UMTRA site)	completed
Pennsylvania	Oak Ridge	Aliquippa Forge (FUSRAP site)	completed
Pennsylvania	Oak Ridge	C.H. Schnoor (FUSRAP site)	completed
Pennsylvania	Albuquerque	Canonsburg (UMTRA site)	completed
Tennessee	Oak Ridge	Elza Gate (FUSRAP site)	completed
Tennessee	Oak Ridge	Oak Ridge Associated Universities (ORAU)	completed
Texas	Albuquerque	Falls City (UMTRA site)	completed
Utah	Albuquerque	Green River (UMTRA site)	completed
Utah	Albuquerque	Mexican Hat (UMTRA site)	completed
Utah	Albuquerque	Salt Lake City (UMTRA site)	completed
Wyoming	Albuquerque	Riverton (UMTRA site)	completed
Wyoming	Albuquerque	Spook (UMTRA site)	completed

Table C.2
Sites Completed in 1997

State	Operations/ Field Office	Site	Completion Date
California	Oakland	Geothermal Test Facility	1997
Colorado	Albuquerque	New Rifle (UMTRA site)	1997
Colorado	Albuquerque	Old Rifle (UMTRA site)	1997
Colorado	Albuquerque	Slick Rock Old North Continent (UMTRA site)	1997
Colorado	Albuquerque	Slick Rock Union Carbide (UMTRA site)	1997
Florida	Albuquerque	Pinellas Plant	1997
Illinois	Chicago	Fermi National Accelerator Laboratory	1997
Illinois	Chicago	Site A	1997
Massachusetts	Oak Ridge	Ventron (FUSRAP site)	1997
New Jersey	Oak Ridge	New Brunswick Site (FUSRAP site)	1997

Table C.3
Sites Transferred to the United States Army Corps of Engineers

State	Operations/ Field Office	Site	Completion Date
Connecticut	Oak Ridge	Combustion Engineering (FUSRAP site)	transferred
Illinois	Oak Ridge	Madison (FUSRAP site)	transferred
Maryland	Oak Ridge	W.R. Grace & Company (FUSRAP site)	transferred
Massachusetts	Oak Ridge	Shpack Landfill (FUSRAP site)	transferred
Missouri	Oak Ridge	Latty Avenue Properties (FUSRAP site)	transferred
Missouri	Oak Ridge	St. Louis Airport Site (FUSRAP site)	transferred
Missouri	Oak Ridge	St. Louis Airport Site (Vicinity Properties) (FUSRAP site)	transferred
Missouri	Oak Ridge	St. Louis Downtown Site (FUSRAP site)	transferred
New Jersey	Oak Ridge	DuPont & Company (FUSRAP site)	transferred
New Jersey	Oak Ridge	Maywood (FUSRAP site)	transferred
New Jersey	Oak Ridge	Middlesex Sampling Plant (FUSRAP site)	transferred
New Jersey	Oak Ridge	Wayne (FUSRAP site)	transferred
New York	Oak Ridge	Ashland 1 (FUSRAP site)	transferred
New York	Oak Ridge	Ashland 2 (FUSRAP site)	transferred
New York	Oak Ridge	Bliss & Laughlin Steel (FUSRAP site)	transferred
New York	Oak Ridge	Colonie (FUSRAP site)	transferred
New York	Oak Ridge	Linde Air Products (FUSRAP site)	transferred
New York	Oak Ridge	Niagara Falls Storage Site (FUSRAP site)	transferred
New York	Oak Ridge	Seaway Industrial Park (FUSRAP site)	transferred
Ohio	Oak Ridge	Luckey (FUSRAP site)	transferred
Ohio	Oak Ridge	Painesville (FUSRAP site)	transferred

Table C.4
Sites with Ongoing Remediation Activities

State	Operations/ Field Office	Site	Completion Date
Alaska	Nevada	Amchitka Island (Nevada Offsite)	2001
California	Albuquerque	Sandia National Laboratories - California	1999
California	Oakland	General Atomics Site	2000
California	Oakland	General Electric Vallecitos Nuclear Center	2005
California	Oakland	Laboratory for Energy - Related Health Research	2002
California	Oakland	Lawrence Berkeley National Laboratory	2003
California	Oakland	Lawrence Livermore National Laboratory Main Site	2006
California	Oakland	Lawrence Livermore National Laboratory Site 300	2006
California	Oakland	Energy Technology Engineering Center (ETEC)	2006
California	Oakland	Stanford Linear Accelerator Center	2000
Colorado	Albuquerque	Grand Junction Office Site	2002
Colorado	Albuquerque	Maybell (UMTRA site)	1998
Colorado	Albuquerque	Naturita (UMTRA site)	1998
Colorado	Nevada	Rio Blanco (Nevada Offsite)	2005
Colorado	Nevada	Rulison (Nevada Offsite)	1998
Colorado	Rocky Flats	Rocky Flats Environmental Technology Site	2010/ 2006[a]
Idaho	Chicago	Argonne National Laboratory - West	2000
Idaho	Idaho	Idaho National Engineering and Environmental Laboratory	2050
Illinois	Chicago	Argonne National Laboratory - East	2002
Iowa	Chicago	Ames Laboratory	1999
Kentucky	Albuquerque	Maxey Flats Disposal Site	2002
Kentucky	Oak Ridge	Paducah Gaseous Diffusion Plant	2010
Mississippi	Nevada	Salmon Site (Nevada Offsite)	1999
Missouri	Albuquerque	Kansas City Plant	1999
Missouri	Oak Ridge	Weldon Spring Site	2002
Nevada	Nevada	Central Nevada Test Site	2006
Nevada	Nevada	Nevada Test Site	2014
Nevada	Nevada	Shoal Site (Nevada Offsite)	2004

Table C.4 (Continued)
Sites with Ongoing Remediation Activities

State	Operations/ Field Office	Site	Completion Date
Nevada	Nevada	Tonopah Test Range Area	2007
New Jersey	Chicago	Princeton Plasma Physics Laboratory	1999
New Mexico	Nevada	Gasbuggy (Nevada Offsite)	2005
New Mexico	Nevada	Gnome-Coach (Nevada Offsite)	2004
New Mexico	Albuquerque	Lovelace Respiratory Research Institute (LRRI)	2000
New Mexico	Albuquerque	Los Alamos National Laboratory	2017
New Mexico	Albuquerque	Sandia National Laboratories - NM	2001
New Mexico	Carlsbad	Waste Isolation Pilot Plant	2038
New York	Chicago	Brookhaven National Laboratory	2006
New York	Oakland	Separations Process Research Unit (SPRU)	2014
New York	Ohio	West Valley Demonstration Project	2005
North Dakota	Albuquerque	Belfield (UMTRA site)	1998
North Dakota	Albuquerque	Bowman (UMTRA site)	1998
Ohio	Ohio	Columbus Environmental Management Project - King Avenue	1998
Ohio	Ohio	Columbus Environmental Management Project - West Jefferson	2005
Ohio	Ohio	Fernald Environmental Management Project	2008/ 2005[b]
Ohio	Ohio	Miamisburg Environmental Management Project	2005[c]
Ohio	Ohio	Ashtabula Environmental Management Project	2003
Ohio	Oak Ridge	Portsmouth Gaseous Diffusion Plant	2005
Puerto Rico	Oak Ridge	Center for Energy and Environmental Research	1998
South Carolina	Savannah River	Savannah River Site	2038
Tennessee	Oak Ridge	Oak Ridge Reservation (Y-12, ORNL, ETTP, ORR)	2013
Texas	Albuquerque	Pantex Plant	2002
Utah	Albuquerque	Monticello Millsite and Vicinity Properties	2001
Washington	Richland	Hanford Site	2046

[a]The Rocky Flats Environmental Technology Site is committed to accelerate activities to complete the site in 2006.
[b]The Ohio Field Office and the Fernald Environmental Management Project are committed to accomplishing completion scheduled for 2008 by the end of 2005.
[c]Pending validation of the current baseline, it is the goal of the Miamisburg Environmental Management Project and the Ohio Field Office to clean up the site by the end of 2003.

Appendix D

Programmatic Risk

The purpose of the programmatic risk concept is to provide each site an opportunity to identify areas of uncertainty (i.e., risk to cost, schedule, and technical performance) associated within the strategy to accelerate site closure dates. As Operations/Field Offices take on the challenge of accelerating site closure, areas with high programmatic risk will become the focus of DOE management attention to insure appropriate visibility and resources are provided. The major objective is to eliminate, as early as possible, those project uncertainties that can result in unexpected growth to cost and schedule. Programmatic risk is associated with a project's cost, schedule, and performance; it should not be confused with risk to the worker, public, and environment.

Each site strategy describes the "critical closure path" for the major activities required for site closure. The critical closure path is a streamlined schedule of high-level activities, events, and/or decisions that must occur "on schedule" to achieve the site closure date. The critical closure path is composed of two sources of schedule information: Critical Path and Critical Events.

A. **Critical Path** information is obtained from the site's analysis of all activities scheduled to complete the EM mission and achieve closure. It is defined as the longest path (in terms of duration) through the schedule of project activities that achieve site closure. The duration of activities on the critical path drives the site closure date. Delay in a critical path activity will delay the closure of the site; similarly, acceleration of the site closure date can occur only if acceleration occurs with critical path activities. Many other non-critical path activities are included in the site's strategy; however, sufficient float (i.e., slack time) exists with these activities to allow some flexibility in their accomplishment without affecting the site closure date.

B. **Critical Events** are those selected milestones, events, decisions, and/or activities that are not on the critical path but are of sufficient programmatic risk to warrant upper-level DOE management and stakeholder attention. Milestones selected to be critical events should be extracted from those included in the site's Project Baseline Summaries.

Programmatic risk categories are described in Table D-1.

Table D-1: Programmatic Risk Categories

Risk Categories	Technological	Work Scope Definition	Intersite Dependency
5 (high)	The technology required to accomplish the planned activity does not exist Development of this technology has not been initiated, but an STCG number has been assigned	Project end state is not determined or supported by stakeholders Waste/material quantities and characteristics are unknown Process operations are not identified or supported by stakeholders Final disposition location for waste/material has not been identified	Activity involves multiple sites No concurrence has been reached between sites Stakeholders are opposed to the site's involvement in the activity
4 (high)	The technology to accomplish the planned activity is identified and has an STCG number Development of the technology is only at the laboratory level	Project end state is determined but may be controversial to stakeholders Process operations are identified but may be controversial to stakeholders Final disposition location for waste/material has not been identified and approved.	Activity involves multiple sites, site concurrence has been verbally reached The Waste Acceptance Criteria (WAC) has not been resolved No funding has been identified and no schedule for receipt or treatment of the waste/material exists Involvement of the site may be controversial to stakeholders
3	The technology required has been identified and has an STCG number assigned Technology is in full scale development and demonstration	Project end state is determined and is expected to be acceptable to stakeholders Waste/material quantities and characteristics are broadly known Process operations are identified and expected to be acceptable to stakeholders Final disposition location for waste/material has been identified and an EIS is being prepared	Activity impacts another site, site concurrence has been verbally reached Receiving facility is reviewing characterization data to determine WAC acceptability Funding has been identified but no schedule for receipt or treatment of the waste/material exists Site involvement is expected to be acceptable to stakeholders

Table D-1: Programmatic Risk Categories (Continued)

Risk Categories	Technological	Work Scope Definition	Intersite Dependency
2	The required technology has been fully developed and demonstrated at another site with a similar waste/material type	Project end state is determined and is expected to be acceptable to stakeholders Waste/material quantities and characteristics are broadly known Process operations are identified and expected to be acceptable to stakeholders Final disposition location for waste/material has been identified and an EIS is being prepared	Activity doesn't impact another site or site concurrence has been documented if multiple sites are impacted Receiving facility has verified WAC acceptability Funding has been identified but no schedule for receipt or treatment of the waste/material exists Site involvement is supported by stakeholders
1 (low)	Technology has been demonstrated at the site on some actual waste/materials and is operationally ready	Project end state is determined and supported by stakeholders Waste/material quantities and characteristics are well known Process operations are identified and supported by stakeholders Final disposition location for waste/material has been identified and an EIS ROD is prepared	Activity doesn't impact another site or site concurrence has been documented if multiple sites are involved Receiving facility has verified WAC acceptability Funding is identified in an approved PBS and facility is ready to receive the waste/material Site involvement is supported by stakeholders

Appendix E

Environmental Management Site Cleanup Summaries

Appendix E presents eight of the Department of Energy's (DOE's) Operations/ Field Office summaries that were not presented in Chapter 3. Each summary contains a discussion of the Office of Environmental Management (EM) mission managed by the Operation/Field Office. The discussion is broken into five sections: a general overview; a discussion of end state assumptions; the cost and completion dates for the sites and projects; a work scope summary; and the critical closure paths and programmatic risks of the strategy managed under the Operations/Field Office.

Included as part of each work scope summary is a "Conceptual Summary Disposition Map." These maps show a summary of each office's current conceptual life-cycle approaches for managing EM wastes, nuclear materials, and contaminated media–from their current status, through storage, treatment, and disposal–to achieve the assumed site end states described in the relevant site strategy. In some cases, these conceptual approaches include shipping and off-site treatment and disposal. The Conceptual Summary Disposition Maps represent a "roll-up" from site-, waste-, material-, and media-specific maps. Volumes are approximate and have been rounded to two significant figures. The maps represent data approved as of February 1998. Since then, EM has carried out an effort to reconcile discrepancies and improve data quality. Although these improvements will not appear in *Paths to Closure* until the next update, they are reflected in the current "working" data set that EM continually updates as sites make changes.

The EM site cleanup summaries are presented in the following order:

- Albuquerque Operations Office
- Carlsbad Area Office
- Chicago Operations Office
- Idaho Operations Office
- Nevada Operations Office
- Oak Ridge Operations Office
- Oakland Operations Office
- Ohio Field Office

Additional information on all of the Operations/Field Offices can be found in the site versions of *Paths to Closure* and other supporting documents.

Conceptual Summary Disposition Maps compile information for the sites that report through the Operations or Field Offices. The maps do not reflect Headquarters-directed or national-level strategies for each site, Operations Office, or Field Office. Within each map, activities are organized into "streams," which are defined as groups of materials, media, or wastes having similar origins, management requirements, or barriers to disposition. The following seven waste, material, and media categories are depicted in the maps:

- High-level waste (HLW)
- Transuranic waste (TRU)
- Mixed low-level waste (MLLW)
- Low-level waste (LLW)
- Environmental restoration activities (ER)
- Spent nuclear fuel (SNF)
- Nuclear materials

As has always been the case for this planning effort (reflected in December 1996 and October 1997 guidance to sites) implementation of each element of the EM program is contingent upon the completion of whatever evaluation is required under the National Environmental Policy Act (NEPA), the Comprehensive Environmental Response, Compensation, and Liability Act (CERCLA), or other statutes.

Decisions that remain to be made include those resulting from two DOE Environmental Impact Statements (EISs). Decisions on disposition of certain nuclear materials will be made pursuant to the Department's *Management of Certain Plutonium Bearing Residues and Scrub Alloys at the Rocky Flats Environmental Technology Site Environmental Impact Statement*. Until these decisions are made, the Conceptual Summary Disposition Maps reflect the "to be decided" (or "TBD") status of those materials.

Decisions on five waste types have been or will be made pursuant to the Department's May 1997 *Final Waste Management Programmatic Environmental Impact Statement* (WM PEIS). This nationwide NEPA analysis examined the potential environmental impacts of managing more than 2 million cubic meters of wastes from past, present, and future DOE activities. The Final WM PEIS identified preferred alternatives for transuranic waste treatment and storage, high-level waste storage, and hazardous waste treatment. The Department has identified preferred management strategies for mixed low-level waste treatment and disposal and low-level waste treatment and disposal. Preferred sites for these management activities have not yet been identified. In this appendix, assumptions regarding low-level and mixed low-level wastes are subject to change based on future Records of Decision (RODs). The Department has committed to publicly identify its preferred sites at least 30 days prior to issuing any ROD for these two waste streams. As of February 1998, one ROD has been issued from the WM PEIS process for transuranic waste treatment and storage. The Conceptual Summary Disposition Maps show specific disposition of transuranic waste, consistent with this ROD.

The Conceptual Summary Disposition Maps' depiction of environmental restoration activities differ from other waste or material management activities. Disposition paths for environmental restoration activities begin with "Contaminated Media" and show a "Response Strategy" for the media. Those strategies may or may not be based on decisions regarding environmental restoration wastes resulting from the CERCLA, NEPA, and Resource Conservation and Recovery Act (RCRA) processes. Where such decisions have not yet been made, environmental restoration planning was based upon assumptions that are being evaluated under CERCLA, NEPA, and/or RCRA, and may change as more media characterization data become available, as comments are received from local stakeholders through public involvement processes, or as the regulatory agencies review and evaluate the various cleanup alternatives.

E.1 Albuquerque Operations Office Summary

The Albuquerque Operations Office is located on Kirtland Air Force Base, directly south of the City of Albuquerque, New Mexico. Historically, the Albuquerque Operations Office's primary mission had been to manage sites that were involved in research, development, production, and maintenance of nuclear weapons. In recent years, this mission has evolved to include environmental management, science and technology, technology transfer and commercialization, and national energy objectives.

Albuquerque Operations Office

The Albuquerque Operations Office provides oversight for environmental management activities to the following sites: the Lovelace Respiratory Research Institute (formerly the Inhalation Toxicology Research Institute); the Los Alamos National Laboratory; the Sandia National Laboratories in New Mexico and California; the South Valley Superfund Site; the Kansas City Plant; the Grand Junction Office; the Pantex Plant; Maxey Flats; the Pinellas Plant; the Monticello Millsite; and uranium mill tailings sites.

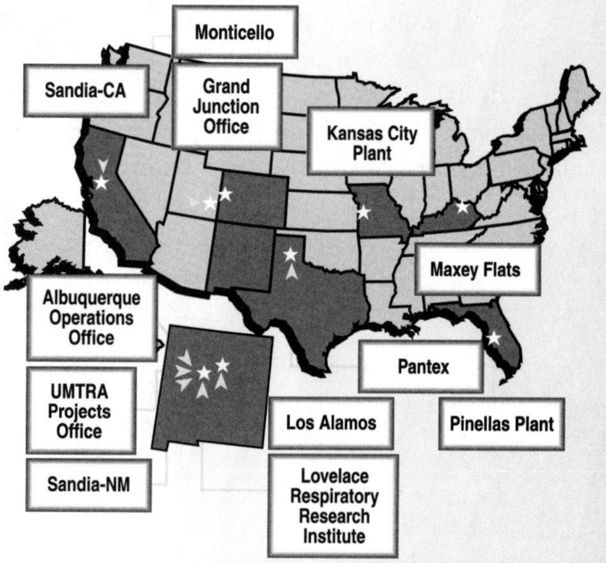

The ***Lovelace Respiratory Research Institute*** (formerly the Inhalation Toxicology Research Institute) was established in 1960 to conduct research on the human health consequences of inhaling airborne radioactive materials. Beginning in the 1980s, the program shifted to more basic research on the human respiratory tract and its response to inhaled toxicants.

Los Alamos National Laboratory was established in 1943 to design, develop, and test nuclear weapons. Research programs in nuclear physics, hydrodynamics, conventional explosives, chemistry, metallurgy, radiochemistry, and life sciences supported this mission. In addition to research, an important function of the Laboratory has been processing plutonium metal and alloys from nitrate solution feedstock provided by other production facilities. Processing plutonium metal took place from 1945 to 1978. Other operations included reprocessing nuclear fuel, processing polonium and actinium, and producing nuclear weapons components.

The ***Sandia National Laboratories*** were established in the 1940s as the engineering arm of the nuclear weapon development program. Sandia National Laboratories - New Mexico is a multi-program national laboratory with research and development programs in a broad range of scientific and technical fields, including fundamental energy research, energy conservation and renewable energy, nuclear reactor safety and reliability, nuclear waste management, and magnetic-confinement fusion. Sandia National Laboratories - California was established in 1956 to conduct research and development in the interest of national security, with principal emphasis on nuclear weapons development and engineering, excluding nuclear materials. It enabled a close working relationship with Lawrence Livermore National Laboratory.

The ***Kansas City Plant*** was constructed in 1942 to build aircraft engines for the Navy. After World War II, it was used for storage, and in 1949 it was selected for its current mission, the manufacturing of nonnuclear components for nuclear weapons. Electrical, electromechanical, mechanical, and plastic components are manufactured or procured by this facility.

Maxey Flats was opened under a lease arrangement between the Commonwealth of Kentucky and the Nuclear Engineering Company (now U.S. Ecology, Inc.) of Louisville, Kentucky in January 1963. The site contains long-lived radionuclides brought to the site from research laboratories, electric utilities, government and private health care facilities, manufacturing companies, and nuclear powerplants throughout the United States. The Department of Energy (DOE) has no management responsibilities for the cleanup of this site, but pays a share of the costs.

The ***Pantex Plant*** was built by the United States Army in 1942 as a conventional bomb plant. The mission of the Pantex Plant involves fabricating high explosives for nuclear weapons, assembling nuclear weapons, maintaining and evaluating nuclear weapons in the stockpile, and dismantling nuclear weapons as they are retired from the stockpile. At present, the principal operation is the disassembly of nuclear weapons.

The ***Pinellas Plant*** has been part of the DOE's nuclear weapons complex since 1957. The plant's former mission was component fabrication. In September 1994, the plant stopped producing weapons-related components and began the transition from a defense mission to an environmental management mission. In 1997 this facility was closed and transferred to Pinellas County.

Grand Junction Office was established in 1943 under the Manhattan Engineer District. Between 1943 and 1946, the U.S. Vanadium Corporation constructed and operated a uranium refinery for the federal government at the site. As a result of past uranium-related activities, surface and near-surface soils, buildings (wood, concrete/brick and metal), and related equipment were contaminated with uranium mill tailings and ore. In addition to the cleanup of this contamination, the Grand Junction Office also serves as a central office for managing long-term surveillance and monitoring at some DOE sites.

Monticello Millsite and Vicinity Properties were transferred to the Department of Energy's Environmental Management (EM) program in 1987 for the remediation of contamination caused by past vanadium and uranium milling at the millsite. The Grand Junction Office is responsible for managing the cleanup activities at the Monticello Millsite.

The ***Uranium Mill Tailings Remedial Action (UMTRA) Surface Projects and UMTRA Groundwater Projects*** manage the implementation of the Uranium Mill Tailing Radiation Control Act (UMTRCA). The United States Congress passed the UMTRCA in 1978 in response to public concern regarding potential health hazards of long-term exposure to radiation from uranium mill tailings. The Act authorized DOE to stabilize, dispose of, and control uranium mill tailings and other contaminated material at 24 uranium mill processing sites and approximately 5,200 associated vicinity properties. The 24 UMTRA sites include: Ambrosia Lake (New Mexico), Belfield (North Dakota), Bowman (North Dakota), Canonsburg (Pennsylvania), Durango (Colorado), Falls City (Texas), Grand Junction (Colorado), Green River (Utah), Gunnison (Colorado), Lakeview (Oregon), Lowman (Idaho), Maybell (Colorado), Mexican Hat (Utah), Monument Valley (Arizona), Naturita (Colorado), New Rifle (Colorado), Old Rifle (Colorado), Riverton (Wyoming), Salt Lake City (Utah), Shiprock (New Mexico), Slick Rock - Old North Continent (Colorado), Slick Rock - Union Carbide (Colorado), Spook (Wyoming), and Tuba City (Arizona).

E.1.1 End State

The Albuquerque Operations Office planned end states for each site at completion are compliance-based and can be achieved with currently available technology. Therefore, they are not likely to be modified as new technologies become available. While economics are likely to affect schedules, the Albuquerque Operations Office does not expect economic feasibility issues to affect planned end states significantly. Unanticipated new regulatory requirements have the greatest potential to change the planned end states at Albuquerque Operations Office sites.

The landlord programs at non-EM sites will have responsibility for determining future use and final end state at the completion of EM activities. Facilities being decontaminated or decommissioned under EM programs will revert to landlord control upon completion. Plans call for EM control of active waste management facilities to be transferred to the generator or landlord program by 1999. While EM activities will terminate, these facilities will continue to operate with the final end state to be determined by the landlord program. Also, at these sites, DOE will maintain stewardship and overall land use will likely continue as is for the foreseeable future. Exhibit E-1 provides a summary of the anticipated end states for sites managed by the Albuquerque Operations Office.

Exhibit E-1

Summary of Albuquerque Operations Office End States

Site Name	End State Description
Ambrosia Lake (completed UMTRA site)	A Nuclear Regulatory Commission (NRC)-licensed disposal cell with a radon barrier cover and surface layer of rock rip-rap for erosion control will remain on site. Under the provisions of the UMTRCA, public access to the disposal cell will be restricted but future land use at the site is undetermined.
Belfield, Bowman (UMTRA sites)	At the request of the State of North Dakota, the Department has revoked the designation of these two sites under UMTRCA. As a result of the revocation, these sites will no longer require remediation under the UMTRCA and DOE will have no long-term stewardship requirements.
Canonsburg, Falls City, Green River, Lakeview, Lowman, Shiprock, Spook (all completed UMTRA sites), Maybell (UMTRA site)	A NRC-licensed disposal cell will remain at each site. Under the provisions of the UMTRCA, public access to the disposal cell will be restricted but future land use at each site is undetermined. Active groundwater remediation is not planned at this time.
Durango, Grand Junction/Cheney Cell (UMTRA sites)	The tailings have been disposed of in off-site disposal cells licensed by the NRC. Under the provisions of the UMTRCA, public access will be restricted but future land use at each site is undetermined. Site assumptions are that groundwater will undergo natural attenuation until the site meets EPA standards.
Grand Junction Office	Under the Grand Junction Office Remedial Action Project (GJORAP), all radiological contamination will be either removed and disposed of off site or the use of supplemental limits (SL) will be selectively applied and approved. The significantly contaminated buildings will be decontaminated or demolished and the remainder of the contaminated buildings will undergo application and approval of SL so that the entire site can be released for unrestricted use. The remaining land and buildings will be transferred to private or other use, with no restrictions. Administrative control of groundwater will continue until it is verified that passive remediation has achieved cleanup goals.
Gunnison (completed UMTRA site)	All contaminated surface materials have been removed from the site and stabilized in a disposal cell licensed by the NRC. Site assumptions are that groundwater will undergo natural attenuation until the site meets EPA standards.

Exhibit E-1 (Continued)

Site Name	End State Description
Lovelace Respiratory Research Institute (LRRI)	This site was cleaned to industrial standards and closed in 1996 with neither surveillance nor monitoring activities required. Contaminated soil was shipped off site, but groundwater contamination exceeds the cleanup level of 10 mg/l set by the New Mexico Environmental Department. Natural attenuation of the nitrates is expected to reduce groundwater contamination levels below the cleanup standard. LRRI is located on land which the U.S. Air Force leases to DOE. DOE's Office of Energy Research is the current operational landlord and will likely make future mission and end state decisions. LRRI will continue to manage DOE generated waste as long as a DOE mission continues.
Kansas City Plant	Soil contamination will be contained or removed by the end of FY 1998. Groundwater contamination, primarily dense non-aqueous phase liquids, will be cleaned up primarily through the use of innovative technologies; however, final contaminant levels are undecided. Groundwater treatment and monitoring is expected to continue from as little as two years to potentially hundreds of years, depending on the outcome of the ongoing negotiations between DOE and EPA. Future land use is expected to be commercial. Defense Programs is the landlord.
Los Alamos National Laboratory	Los Alamos has an ongoing research mission. Legacy mixed low-level waste will be sent off site by 2004. Decommissioning and decontamination of the two on-site TRU reduction and repackaging facilities will be complete by FY 2017. The site will maintain most of its 43 square mile property but is considering transfer of up to 7,000 acres to the county for industrial use. Land and facilities that DOE will retain will be remediated to allow for industrial use. The land that has been released or is scheduled to be released will be remediated to allow for unrestricted use. The Los Alamos environmental restoration project will be complete by 2008.
Maxey Flats Disposal Site	In accordance with the CERCLA ROD, planned cleanup levels will result in natural stabilization with waste remaining on site. DOE has no control or management responsibility. There is no further DOE liability after DOE makes its final payment, currently scheduled for 2001. The Commonwealth of Kentucky is responsible for long-term stewardship. The site will remain a permanent low-level waste disposal site, and will be under controlled access.

Exhibit E-1 (Continued)

Site Name	End State Description
Monticello Millsite & Vicinity Properties	DOE-owned land on the mill site is expected to be deeded to the City of Monticello for recreational use. The Monticello Mill Tailings Site and the Monticello Vicinity Properties Site will be remediated to the radium-226 standards established in 40 CFR 192. Tailings and tailings-contaminated soil will be excavated and placed in a permanent repository on DOE-owned property. A cover will be placed over the tailings to control radon emissions, infiltration of precipitation, and erosion. EPA and the State have approved supplemental standards, with some qualifications, for some vicinity and peripheral properties. Areas that meet radium-226 standards will be released for unrestricted use. Final land use restrictions for other areas are being determined by DOE, EPA, and the State. The on-site repository will remain under DOE control. The remedy for contaminated sediment, surface water, and groundwater has not yet been selected.
Monument Valley (completed UMTRA site)	Surface materials have been shipped to the Mexican Hat UMTRA site for disposal. Site assumptions are that groundwater at Monument Valley will undergo active remediation through 2010 in order to meet EPA groundwater standards.
Naturita (UMTRA site)	All buildings at the site have been demolished. Residual radioactive surface materials have been transported to the Uravan disposal site and disposed of in a disposal cell at the Upper Burbank Repository. Site assumptions are that groundwater will undergo natural attenuation until the site meets EPA standards.
New Rifle Site, Old Rifle Site (UMTRA sites)	Surface materials have been excavated, transported, and disposed of at the Estes Gulch disposal cell. Groundwater will undergo natural attenuation until the site meets EPA standards. It is expected that the State of Colorado will transfer ownership to the city or county for public use with restrictions; this will allow DOE access to continue the UMTRA groundwater project.
Pantex Plant	Site closure under the Environmental Management program is not anticipated in the foreseeable future. As a result, facility decontamination and decommissioning and future land use are not addressed in *Paths to Closure*. Current land use (industrial) will remain unchanged. Waste management operations will continue in support of the site's ongoing mission. Legacy waste will be dispositioned by FY 2004. All currently identified release sites will be remediated to achieve closure designation in accordance with cleanup levels contained in the Texas Risk Reduction Standards Guidance. Groundwater pump and treat operations will continue until FY 2015. However, long-term efficiency and capability of the groundwater extraction and treatment system to capture the contaminant plume is uncertain, and additional time could be required to fully achieve groundwater remediation objectives.

Exhibit E-1 (Continued)

Site Name	End State Description
Pinellas Plant	This site was sold to Pinellas County Industrial Council (PCIC) in FY 1995, and DOE completed surface remediation in FY 1997. Pinellas' liability under CERCLA for former off-site waste disposal was transferred to the Grand Junction Office as of October 1997. The site will require treatment of contaminated groundwater where high levels of groundwater contamination exist to meet the "industrial with unrestricted access" classification. Groundwater will be cleaned to Clean Water Act maximum contaminant levels. When site groundwater is remediated to the specified level, DOE's responsibility will be terminated.
Riverton (completed UMTRA site)	Site assumptions are that groundwater at Riverton has been determined a non-drinking water source and will undergo natural attenuation until the site meets EPA standards (up to 100 years).
Salt Lake City (completed UMTRA site)	Tailings have been shipped off site for disposal. The site remains under private control. Current planning is that Clean Water Act alternate concentration limits will be accepted for achieving groundwater compliance.
Sandia National Laboratories - California	Sandia will have an ongoing mission under the responsibility of the Office of Defense Programs. The Sandia Environmental Restoration Project intends to complete remediation and associated waste disposal for all 23 release sites by 1999. All designated solid waste management units and areas of concern will be remediated or placed under management controls such that no further action is necessary. The Environmental Restoration Project is planning to close the Navy Landfill in 1998.
Sandia National Laboratories - New Mexico	This site will have an ongoing mission under the responsibility of the Office of Defense Programs. All identified environmental restoration sites will have been remediated and associated waste disposed of in a Corrective Action Management Unit (CAMU) disposal cell or at an off-site location. All 183 sites except the chemical waste landfill, mixed waste landfill, and the CAMU disposal cell will be released for reapplication by Defense Programs. By 2001 disposal of all historical waste, waste generated within permit regulatory limits, and closure of excess waste management facilities will be complete. Nearly all of the land is expected to be available for reapplication for DOE/SNL programmatic uses (industrial) beginning in 2001, with security safeguards remaining in place. Some future land use may include recreational activities, although there will be controlled access for the landfills and CAMU.

Exhibit E-1 (Continued)

Site Name	End State Description
Slick Rock - Old North Continent and Union Carbide (completed UMTRA sites)	A NRC-licensed disposal cell with a radon barrier cover and surface layer of rock rip-rap for erosion control has been constructed at an off-site location. Under the provisions of the UMTRCA, public access to the disposal cell will be restricted, but future land use at the site is undetermined. Tailings from both sites have been relocated to an off-site disposal cell. Site assumptions are that groundwater at Old North Continent and Union Carbide has been determined a non-drinking water source and will undergo natural attenuation until the site meets EPA standards (up to 100 years). Albuquerque Operations Office assumes that NRC will complete licensing review by 1999. The sites will be returned to their owners upon NRC certification of compliance with Subpart B of the EPA groundwater protection standards.
South Valley Superfund Site	The surface remediation of this site was completed in 1996. Groundwater contamination continues to threaten local drinking water supplies and private wells. Remediation includes removing the contamination from the groundwater and preventing migration of contamination. Groundwater remediation will take place until eight consecutive groundwater samples indicate all cleanup levels have been achieved or a waiver of technical impracticability is approved by the EPA. DOE, the U.S. Air Force, and General Electric entered into a settlement agreement to reimburse General Electric for environmental restoration services performed at the site.
Tuba City (completed UMTRA site)	A NRC-licensed disposal cell with a radon barrier cover and surface layer of rock rip-rap for erosion control will remain on site. Under the provisions of the UMTRCA, public access to the disposal cell will be restricted. Site assumptions are that groundwater at Tuba City will undergo active remediation through 2010 or beyond in order to meet EPA groundwater standards.

E.1.2 Cost and Completion Dates

The Albuquerque Operations Office has divided its environmental management work into 20 discrete projects including the two Uranium Mill Tailings Remedial Action (UMTRA) projects (one for surface tailings and one for groundwater.) A Project Baseline Summary (PBS) exists for each project and contains detailed information, including cost, schedule, scope, end state, and interim milestones. A summary of the Albuquerque cost and schedule information is illustrated in Exhibit E-2. For additional information about these projects, refer to individual PBSs.

Exhibit E-2 Albuquerque Operations Office
Cleanup Project Summary: Duration and Costs (All costs in thousands of 1998 dollars)

Site Closure Project Activities	1997 - 2006	2007 - 2070	Total
Albuquerque Operations Office			
Albuquerque Miscellaneous Programs	31,227	0	31,227
New Mexico Agreement in Principle (AIP)	15,044	31,771	46,815
Grand Junction			
All Other Projects	150,473	1,209,113	1,359,586
Lovelace Respiratory Research Institute (LRRI)			
LRRI	6,134	10,530	16,664
Kansas City Plant			
Kansas City Plant Environmental Restoration	17,310	66,022	83,332
Los Alamos National Laboratory (LANL)			
Nuclear Material Facility Stabilization R&D	132,155	0	132,155
LANL Environmental Restoration	603,499	193,156	796,655
LANL Waste Management - Legacy Waste	318,663	267,410	586,073
LANL Waste Management - Newly Generated Waste	63,129	0	63,129
Maxey Flats			
Maxey Flats Field Management Project	12,540	0	12,540
Monticello			
Monticello Projects	129,210	0	129,210

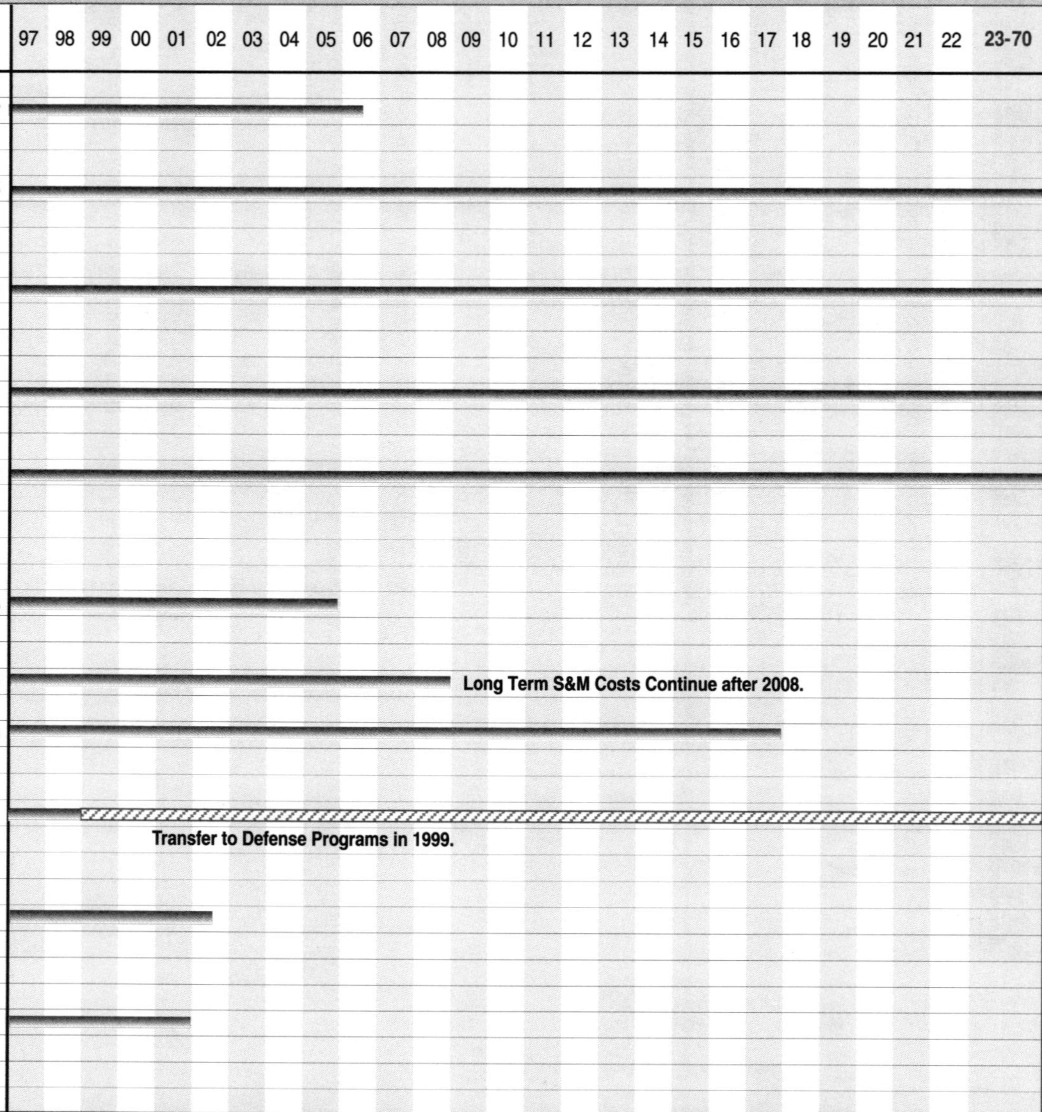

Exhibit E-2 Albuquerque Operations Office (Continued)
Cleanup Project Summary: Duration and Costs (All costs in thousands of 1998 dollars)

Site Closure Project Activities	1997 - 2006	2007 - 2070	Total
Pantex Plant			
Pantex Plant Site Remediation Project	71,917	11,784	83,701
Pantex Waste Operations	28,418	0	28,418
Pinellas Plant			
Pinellas Plant Close-out	102,831	126,003	228,834
Groundwater Cleanup	20,287	13,507	33,794
Sandia National Laboratory (SNL)			
Sandia Environmental Restoration Project	100,714	3,414	104,128
Sandia Waste Management	36,681	0	36,681
South Valley			
South Valley Superfund Site	5,545	1,881	7,426
Uranium Mill Tailings Remedial Actions (UMTRA)			
UMTRA Groundwater	114,669	45,730	160,399
UMTRA Surface Remedial Action Project	134,525	0	134,525
Total	**2,094,971**	**1,980,321**	**4,075,292**

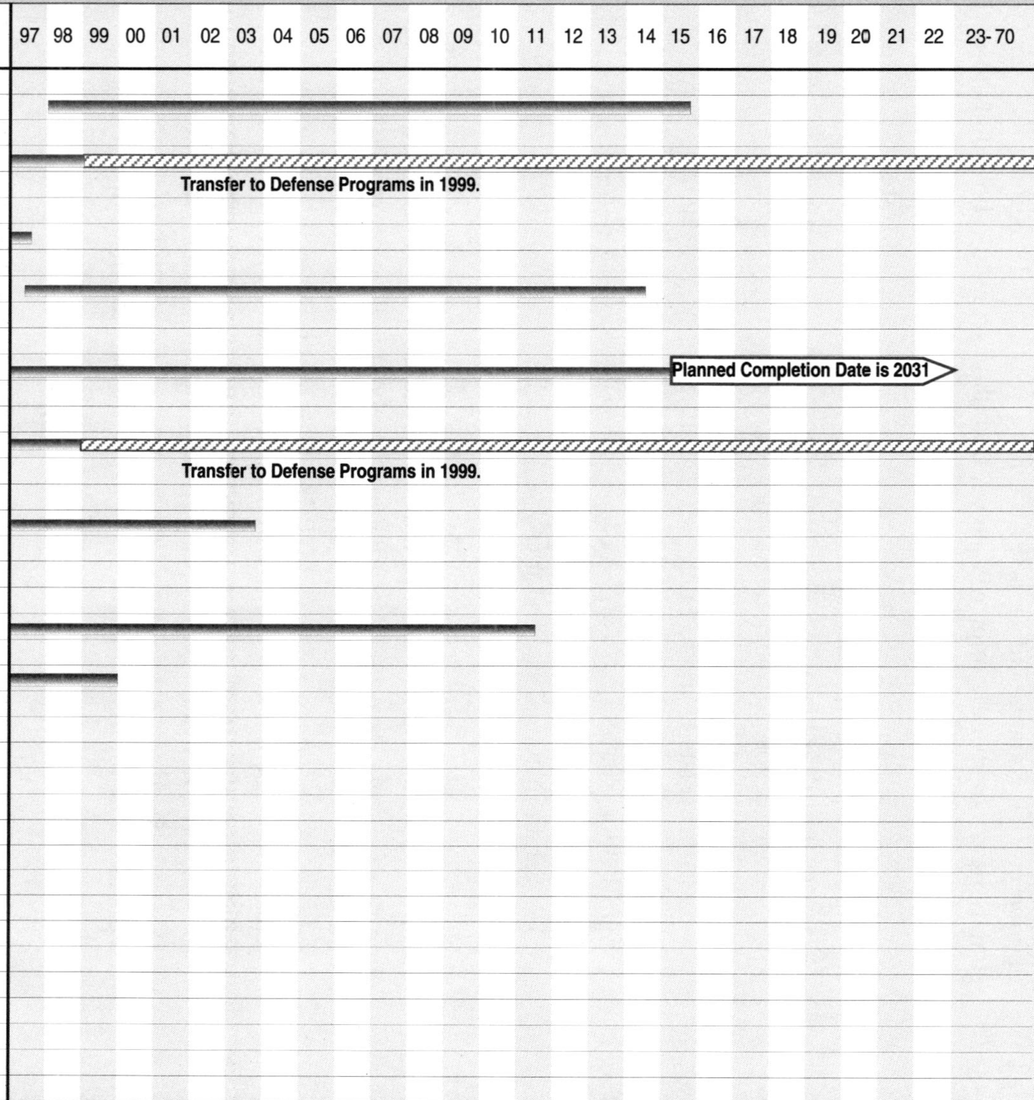

The estimated total EM life-cycle cost of cleanup of the sites managed by the Albuquerque Operations Office is $4.1 billion (constant 1998 dollars). This estimate does not include approximately $4.5 billion (constant 1998 dollars) of non-EM costs. The overall site planned completion dates are as follows:

Site	Date
Grand Junction Office Site	2002
Kansas City Plant	1999
Los Alamos National Laboratory	2017
Lovelace Respiratory Research Institute	2000
Maxey Flats Disposal Site	2002
Monticello Millsite and Vicinity Properties	2001
Pantex Plant	2002
Pinellas Plant	1997
Sandia National Laboratories - CA	1999
Sandia National Laboratories - NM	2001

Within the UMTRA Surface Project, tailings remediation has been completed at 20 processing sites. Two sites (Naturita and Maybell) will be completed in 1998. At the request of the State of North Dakota, DOE has revoked the designation of Belfield and Bowman under UMTRCA. Sites requiring active groundwater remediation will be retained in the UMTRA Groundwater Project until FY 2011, at which time they will be transferred to the long-term surveillance and monitoring program managed by the Grand Junction Office. Presently, three sites are proposed for active remediation, nine sites are proposed for passive remediation, and the remaining 10 sites are proposed for no action.

The projected cost profile for environmental management associated with the Albuquerque Operations Office is developed by combining the cost estimates in each of the PBSs. Exhibit E-3 displays the resultant baseline cost profile.

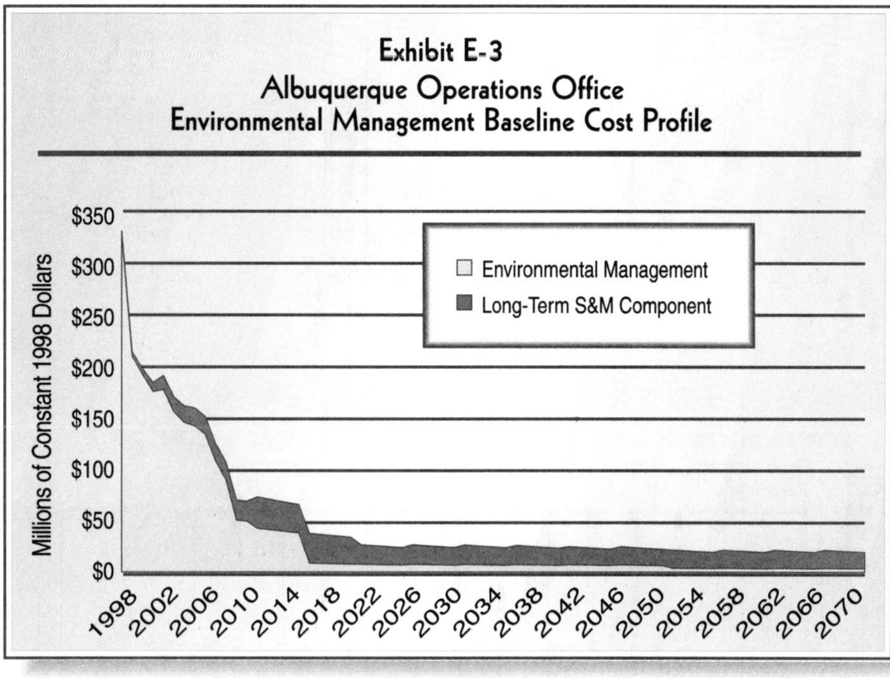

E.1.3 Work Scope Summary

The EM cleanup mission at Albuquerque focuses on the safe and efficient cleanup of national laboratories and production plants within its complex. The scope of work at Albuquerque consists of projects at numerous sites, including the Albuquerque Operations Office, the Lovelace Respiratory Research Institute, the Los Alamos National Laboratory, the Sandia National Laboratories, the South Valley Superfund Site, the Kansas City Plant, the Grand Junction Office, the Pantex Plant, the Maxey Flats Disposal Site, the Pinellas Plant, the Monticello Millsite, and the UMTRA sites. Cleanup activities include the management of groundwater contaminated with residual radioactive materials at UMTRA sites, disposal of low-level waste at Los Alamos National Laboratory, and the disposal of soils and sediments contaminated with radioactive residual materials at the Monticello Mill site. The sections below describe the major waste, material, and contaminated media volumes to be addressed by the Albuquerque Operations Office. The volumes reported are approximate, and correspond to the major waste, material, and media flows, potential treatment processes, and off-site disposal destinations presented in Exhibit E-4, the Albuquerque Operations Office Conceptual Summary Disposition Map.

Exhibit E-4 Albuquerque Operations Office Conceptual Summary Disposition Map

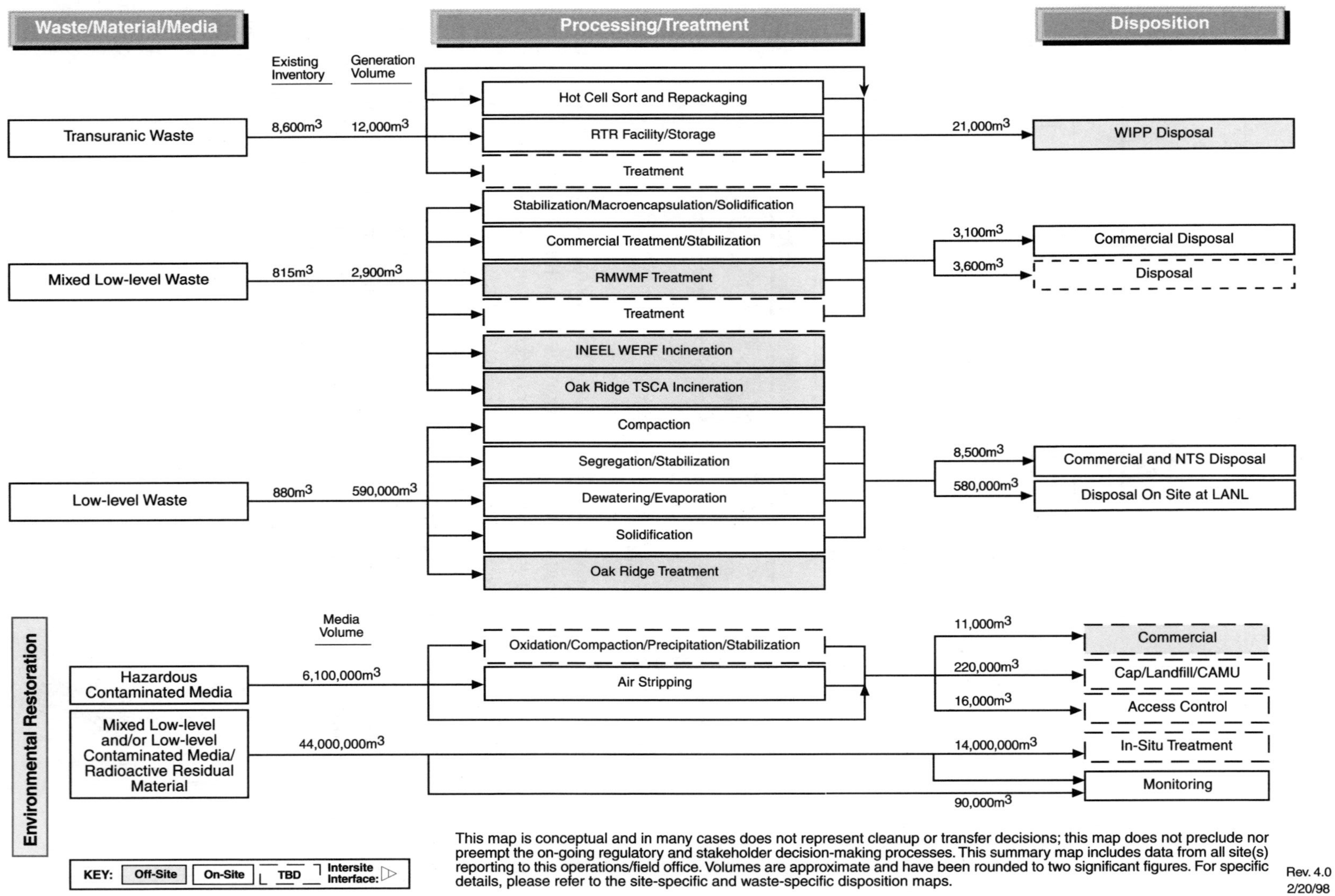

This map is conceptual and in many cases does not represent cleanup or transfer decisions; this map does not preclude nor preempt the on-going regulatory and stakeholder decision-making processes. This summary map includes data from all site(s) reporting to this operations/field office. Volumes are approximate and have been rounded to two significant figures. For specific details, please refer to the site-specific and waste-specific disposition maps.

Rev. 4.0
2/20/98

Transuranic Waste

- Approximately 8,600 cubic meters of transuranic waste are currently in inventory and 12,000 cubic meters are expected to be generated over the life cycle of cleanup operations. After sorting, repackaging, and some treatment, 21,000 cubic meters are expected to be disposed of at the Waste Isolation Pilot Plant (WIPP).

Other Waste

- Approximately 815 cubic meters of mixed low-level waste are currently in inventory, and 2,900 cubic meters are expected to be generated over the life cycle of operations. These waste volumes will be subject to a range of different treatment options, including incineration at DOE sites. After treatment, 3,100 cubic meters are expected to be disposed of at an off-site commercial facility, and an additional 3,600 cubic meters are expected to be disposed of at an off-site location to be determined later.

- Approximately 880 cubic meters of low-level waste are currently in inventory and over 590,000 cubic meters are expected to be generated over the life cycle of operations. Waste volumes will be subject to a range of treatment and processing activities, including transfer to the Oak Ridge Reservation for treatment. After treatment, 8,500 cubic meters are expected to be disposed of at the Nevada Test Site and an off-site commercial facility, and an additional 580,000 cubic meters are expected to be disposed of at the Los Alamos National Laboratory.

Remedial Action and Facility D&D

- Approximately 6.1 million cubic meters of environmental media, including groundwater, soils, and sediments contaminated with hazardous substances will be managed. Some of this media will be subject to a range of treatment activities, while other waste streams will be disposed of directly. Approximately 11,000 cubic meters are expected to be sent to an off-site commercial facility, 220,000 cubic meters are expected to be either capped in place or disposed of in an on-site facility, and 16,000 cubic meters are expected to be subject to access control.

- Approximately 44 million cubic meters of environmental media including groundwater and soil contaminated with radionuclides and hazardous substances will be managed. Approximately 90,000 cubic meters of environmental media will be subject to monitoring and 14 million cubic meters of groundwater are expected to be treated in-situ.

The sum of life-cycle costs at the Albuquerque sites is illustrated in Exhibit E-5, broken out by major work scope category.

Exhibit E-5
Albuquerque Operations Office
Life-Cycle Costs by Category

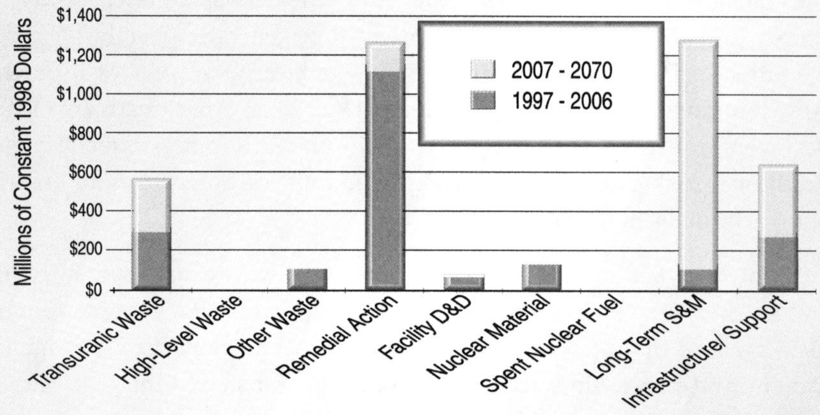

The primary tasks at the Albuquerque sites involve the assessment and remediation of inactive/surplus facilities and contaminated sites; the treatment, storage, and disposal of transuranic, hazardous, and low-level wastes; and the surveillance, environmental monitoring, maintenance, site security, and emergency response for completed environmental cleanup sites from various programs.

E.1.4 Critical Closure Path and Programmatic Risk

The critical closure path schedule, presented as Exhibit E-6, sets forth the timetable for completing closure activities at Albuquerque Operations Office. In the exhibit, the bars represent critical activities. The Albuquerque Operations Office critical closure path reflects those cleanup activities, excluding long-term surveillance and monitoring, which are key to achieving completion of the sites cleanup mission and end states.

Completion of the EM mission at the Albuquerque Operations Office as scheduled will depend on the timely accomplishment of critical activities and milestones. Sites have assigned programmatic risk scores to each of the critical activities/milestones. Appendix D provides a complete definition of programmatic risk. Exhibit E-7 presents a summary of activities and milestones on the critical closure path that have high programmatic risk (programmatic risk scores of 4 or 5 in any category). The Albuquerque Operations Office version of *Paths to Closure* provides more details on the management approach for these high programmatic risk issues.

Exhibit E-6 Albuquerque Environmental Management Critical Closure Path

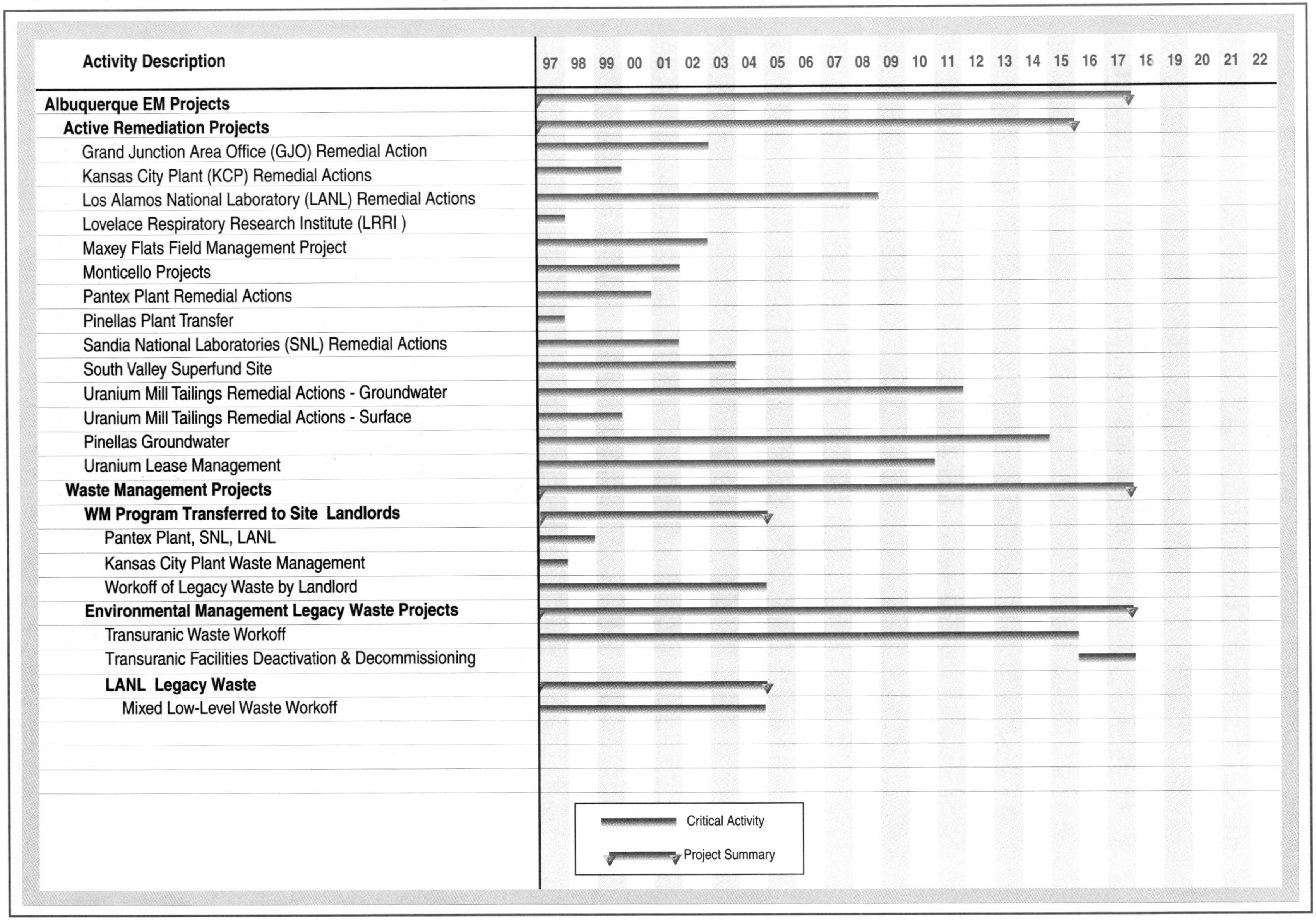

Exhibit E-7

Summary of High Programmatic Risk Activities/Milestones: Albuquerque Operations Office

Site	Project, Activity, Event	Start/End Dates	Programmatic Risk Categories		
			Technological	Work Scope Definition	Intersite Dependency
GJO	Supplemental limits must be approved on Bldg. 7	Oct 99/ Oct 00	1	5	1
	Supplemental limits must be approved on Bldg. 20	Oct 99/ Oct 00	1	5	1
	Buried utilities must be investigated and any contamination found must be minimal	Oct 01/ Sep 02	1	5	3
KCP	Complete 95th Terrace Assessment	Jan 98/ Jun 99	1	4	1
SNL	WIPP opens in May 1998	May 98	1	4	4
	Operation of Auxiliary Hot Cell Facility for packaging of remote-handled TRU	Oct 97/ Sep 99	1	4	4
	Develop system to ship remote-handled TRU to WIPP or Los Alamos National Laboratory	Oct 99/ Sep 00	2	4	5
	Complete mixed waste treatment per Site Treatment Plan (except for mixed TRU waste)	Oct 97/ Sep 02	4	4	5
	Work off of historical MW	Oct 97/ Sep 06	3	4	5
South Valley	Long-term buyout	Apr 03/ Sep 03	1	4	1
UMTRA	Maybell site construction completion	Sep 98	3	4	5
	Maybell site certification license and transfer to GJO for long-term surveillance and monitoring	Sep 99	5	5	5

E.2 Carlsbad Area Office Summary

The mission of the Carlsbad Area Office (CAO) is to protect human health and the environment by opening and operating the Waste Isolation Pilot Plant (WIPP) for safe disposal of transuranic (TRU) waste and by establishing an effective system for management of TRU waste from generation to disposal. It includes personnel assigned to CAO, WIPP site operations, transportation, and other activities associated with the National TRU Program (NTP). The CAO develops and directs implementation of the TRU waste program, and assesses compliance with the program guidance, as well as the commonality of activities and assumptions among all TRU waste sites.

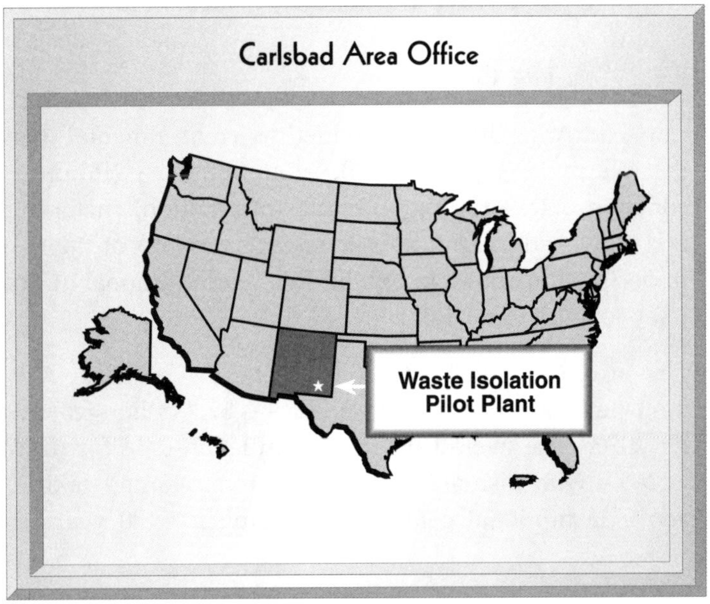

A cornerstone of the Department of Energy's (DOE) national cleanup strategy, WIPP is designed to permanently dispose of TRU waste generated by defense-related activities. Located in southeastern New Mexico, 26 miles east of Carlsbad, project facilities include disposal rooms excavated 2,150 feet underground (about a half mile) in an ancient, stable salt formation. TRU waste consists primarily of tools, gloves, clothing and other such items contaminated with trace amounts of radioactive elements, mostly plutonium. WIPP is scheduled to begin disposing of defense-generated TRU waste in FY 1998. On May 13, 1998, the Secretary of Energy made the decision that WIPP is ready to begin disposal operations after the 30-day Congressionally mandated notification period. However, transportation of TRU waste will be limited to non-mixed waste until the State of New Mexico has issued a Resource Conservation and Recovery Act (RCRA) Part B Permit.

E.2.1 End State

WIPP is neither a "cleanup" nor "closure" site. It is the only TRU waste disposal site in the world. TRU waste management activities for both contact-handled (CH) and remote-handled (RH) TRU wastes are projected to be completed by FY 2039 after completing the Disposal Phase in FY 2033, five years for decommissioning of the surface facilities, and permanently closing the underground. In accordance with the Land Withdrawal Amendment Act of 1996

(LWAA), DOE will have disposed of 175,600 cubic meters of TRU waste in WIPP. Starting in FY 2039, a reduced federal staff and technical contractor support will maintain records of WIPP and the active institutional controls associated with the land withdrawal. Monuments and markers will be built at the site to warn people of the presence of radioactive waste. Active institutional controls over the site will be maintained for 100 years. Low risk has been assigned to this project based upon performance assessments included in the permitting of the facility, which requires no migration of hazardous or radioactive material for 10,000 years. Following completion of the project, there will be no access to the underground. The surface area will be unrestricted for recreational and agricultural uses with the exception of 124 acres which constitute the exclusive-use passive institutional control area.

E.2.2 Cost and Completion Dates

Carlsbad Area Office has divided its environmental management work into five discrete projects. A Project Baseline Summary (PBS) exists for each project and contains detailed programmatic information, including cost, schedule, scope, end state, and interim milestones. A summary of the cost and schedule for these projects is illustrated in Exhibit E-8. For additional information on these projects, refer to individual PBSs.

The estimated EM life-cycle cost of Carlsbad Area Office's TRU waste management and disposal activities is $7.7 billion (constant 1998 dollars) through FY 2070. The overall planned completion date for disposal operations at WIPP is 2033, with dismantling and decommissioning taking another five years and active institutional controls continuing for 100 years thereafter.

Exhibit E-8 Carlsbad Area Office
Cleanup Project Summary: Duration and Costs (All costs in thousands of 1998 dollars)

Site Closure Project Activities	1997 - 2006	2007 - 2070	Total	97-99	00-02	03-05	06-08	09-11	12-14	15-17	18-20	21-23	24-26	27-29	30-32	33-35	36-38
Waste Isolation Pilot Plant (WIPP)																	
WIPP Transuranic Waste Transportation Privatization	40,090	0	40,090														
WIPP Transportation	226,954	716,210	943,164														
WIPP Disposal Phase Certification and Experimental Program	309,522	769,627	1,079,148											Planned Completion Date is 2039			
WIPP Base Operations	1,044,356	3,600,106	4,644,462											Planned Completion Date is 2039			
WIPP Transuranic Waste Sites Integration and Preparation*	204,369	811,606	1,015,975											Planned Completion Date is 2070			
Total	**1,825,290**	**5,897,548**	**7,722,838**														

*WIPP Land Withdrawal Act requires an Active Institutional Control period of 100 years after the WIPP site is dismantled and decommissioned. These activities are reflected in the "WIPP TRU Waste Sites Integration and Preparation" project. The "2007 to 2070" column and the "Total" column reflect costs through FY 2070.

The projected cost profile for environmental management associated with the Carlsbad Office is developed by combining the cost estimates in each of the PBSs. Exhibit E-9 displays the resultant baseline cost profile.

Exhibit E-9
Carlsbad Area Office
Environmental Management Baseline Cost Profile

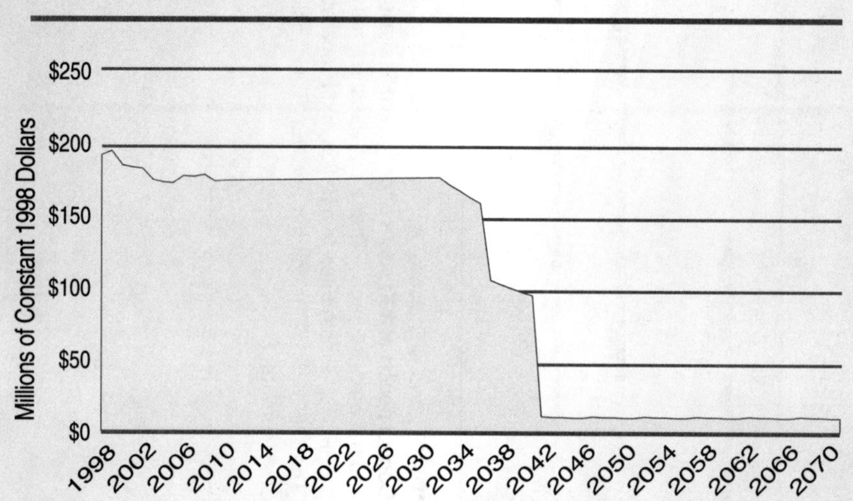

E.2.3 Work Scope Summary

The EM mission at Carlsbad consists of the following work scope.

- The operation of the TRU waste disposal facility which includes all activities required to maintain waste receipt and disposal operations including mining, waste handling and facility operations. Also included in this project are activities required to maintain and operate WIPP that are not directly related to waste disposal.
- The five year recertification cycle of the scientific performance of the facility by the EPA which includes all of the Managing and Operating (M&O), Scientific Advisor and supporting laboratories' experimental, compliance, and performance assessment work in support of certification and operational performance improvement for the WIPP site and the national TRU system. The scope also includes the establishment of a focused international nuclear waste disposal research development program.
- The TRU waste transportation system development and operations This scope includes all site activities required to meet the National TRU Waste Management Plan (NTWMP), Rev. 1, associated with the maintenance and operations of a transportation system. These activities include: emergency

response training; establishing and opening transportation corridors; Ch-TRU and RH-TRU waste packaging initiatives; carrier services; and stakeholder interfaces related to transportation.

The primary locations where TRU waste is currently stored are: Idaho National Engineering and Environmental Laboratory (INEEL), Los Alamos National Laboratory (LANL), Rocky Flats Environmental Technology Site (RFETS), Oak Ridge National Laboratory (ORNL), Savannah River Site (SRS), Hanford Reservation (Hanford), Nevada Test Site (NTS), Lawrence Livermore National Laboratory (LLNL), Argonne National Laboratory - East (ANL-E), and the Miamisburg Environmental Management Project (Mound). Other sites have small quantities of TRU waste that will be disposed of at WIPP. The TRU waste sites scheduled to initially ship CH-TRU waste to WIPP in FY 1998, are INEEL, LANL, and RFETS. Using the shipment schedules in the NTWMP, Hanford, ANL-E, Mound, SRS, and selected small quantity sites will begin shipping waste to WIPP in FY 1999, while LLNL and NTS will begin shipments in FY 2000. By FY 2000, the WIPP facility will be at a full throughput rate of 17 CH shipments per week. In FY 2003, CAO will begin receiving shipments of RH-TRU waste from ORNL and LANL at a rate of two shipments per week and work up to 10 shipments per week by FY 2004.

The process of opening transportation corridors includes cooperative agreements with all Native American tribes along each corridor, state emergency response training, and agreements with the Western Governor's Association and the Southern States Energy Board. CAO also coordinates transportation schedules and plans through the National Governor's Association.

CAO must open and maintain transportation corridors across the United States between each TRU waste site and WIPP. Currently, one corridor from INEEL, RFETS, and LANL is open. Activities to open other corridors require approximately two years prior to shipment campaigns beginning at the sites. The phasing of corridors corresponds with site shipping schedules and eliminates the need for corridor maintenance thus reducing TRU waste complex costs.

- The management activities necessary to direct and integrate the Department's National TRU waste sites activities from generation to disposal including all quality assurances oversight activities This scope includes ongoing TRU integration activities and programs which are directed by the CAO civilian work force. The CAO is the lead office for the management, planning, and integration of the integration of the TRU waste program .

E.3 Chicago Operations Office Summary

The Chicago Operations Office, located at the Argonne National Laboratory site in Illinois, is responsible for the safe and efficient cleanup of national laboratories and other sites under its management. Laboratories managed by the Chicago Operations Office have primary missions relating to energy, nuclear, basic fusion, and high-energy physics research.

Chicago Operations Office

The Chicago Operations Office currently manages environmental restoration and waste management at Ames Laboratory, Argonne National Laboratory-East and West, Brookhaven National Laboratory, Fermi National Accelerator Laboratory, and Princeton Plasma Physics Laboratory.

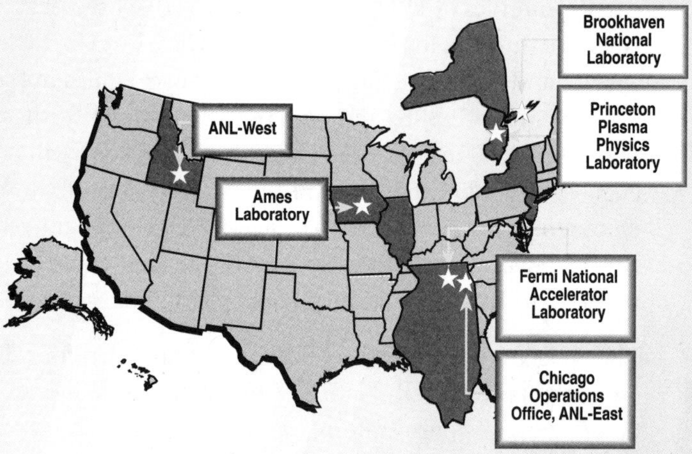

Ames Laboratory was established in the 1940s to develop efficient uranium production processes for the Manhattan Project. The Laboratory's programs now emphasize research in the preparation, characterization, and evaluation of properties of metals and their alloys, especially rare earth metals.

Argonne National Laboratory* - *East (ANL-E) has been involved in research and development activities in support of the Department of Energy (DOE) and its predecessor agencies since 1943. Currently, it serves as a multi-disciplinary research and development laboratory that conducts basic and applied research to support the development of energy-related technologies.

Argonne National Laboratory* - *West's (ANL-W) primary mission was to support liquid metal reactor research and development for the Integral Fast Reactor Program until the program was terminated. Activities at the laboratory now include technology development for spent nuclear fuel and waste treatment, reactor and fuel cycle safety, and facility decommissioning.

Brookhaven National Laboratory has been involved in research and development activities in support of DOE and its predecessor agencies since 1947.

Its current mission is to conduct fundamental research, including conception, design, construction, and operation of large, complex research facilities to carry out both basic and applied research in high energy and nuclear physics.

Fermi National Accelerator Laboratory began its mission as a single-program research and development facility for the Atomic Energy Commission in 1972, when the first accelerator at the laboratory began operations. The laboratory's current mission is to conduct research in high-energy physics under the direction of DOE's Office of Energy Research.

***Princeton Plasma Physics Laboratory* (PPPL)** has historically provided research and development for DOE's fusion energy programs. Currently, activities at the site are devoted to the research and development of plasma fusion energy.

E.3.1 End State

The end state for Environmental Management (EM) program activities at all Chicago sites is completion of all environmental restoration activities by 2006 or sooner and transfer of all waste management activities to the Office of Energy Research, which has landlord responsibilities at the Chicago sites, by FY 2000. All landlord site stewardship and future land use issues will be managed by the Office of Energy Research, with the exception of Argonne National Laboratory-West which will be managed by the Office of Nuclear Energy. Exhibit E-10 provides a summary of anticipated end states for the sites managed by the Chicago Operations Office. In addition to the sites discussed in Exhibit E-10, the Chicago Operations Office supported surveillance and monitoring activities at Site A/Plot M, the Hallam Nuclear Power Facility, and the Piqua Nuclear Power Facility. Those activities will be transferred to the Grand

Exhibit E-10

Summary of Chicago Operations Office End States

Site Name	End State Description
Ames Laboratory	Environmental Restoration will complete its mission in FY 1998 and the Waste Management program is planned to be transferred to Energy Research in 2000. The wastes from the former Chemical Waste Disposal Site, which accepted radiological and chemical wastes, were removed in FY 1995. All of Ames's waste is treated and/or disposed of off site.
Argonne National Laboratory - East	ANL-E will have an ongoing mission, with Energy Research acting as the landlord. The Waste Management Program is planned to be transferred to Energy Research in FY 2000. Corrective action for some release sites will require on-site containment of residual contamination. ANL-E hopes to bring the surplus reactor and nuclear support facilities to meet Nuclear Regulatory Commission unrestricted use standards and remove all postings and warnings by 2002. The majority of work will be complete in 2000.

Exhibit E-10 (Continued)

Site Name	End State Description
Argonne National Laboratory - West	ANL-W has an ongoing mission, and the land is expected to be used for industrial/commercial operations. The Waste Management program was transferred to Nuclear Energy in early FY 1998. Remediation of eight release sites and one facility is in progress. The Central Liquid Processing Area will be decontaminated and decommissioned in FY 1998. Groundwater remediation will be ongoing. The site will become the responsibility of Nuclear Energy in FY 2000.
Brookhaven National Laboratory	Energy Research is the landlord for Brookhaven's ongoing research mission. The Waste Management Program is planned to be transferred to Energy Research in FY 2000. By 2006, soil remediation will be complete, the Boneyard wastes will be disposed of off site, and long-term monitoring will be in place. The groundwater remediation system will be operational. Decontamination and decommissioning of the graphite reactor will be complete. The reactor will be safely and permanently closed, but the final end state for the reactor is not yet defined. Three former on-site landfills have been capped, and one is currently being reused for recreational purposes. Any wastes generated as part of an ongoing mission will be disposed of off site.
Fermi National Accelerator Laboratory	As of the end of FY 1997, EM has no further obligations to Fermi. Funding for managing waste activities at Fermi was transferred to Energy Research in the beginning of FY 1998. All waste is sent off site for appropriate treatment and disposal, as required. As long as Fermi Laboratory is in operation, waste management will be a necessary program function.
Princeton Plasma Physics Laboratory	PPPL will continue to conduct research, and generate of hazardous waste. The Waste Management program is planned to be transferred to Energy Research in FY 2000. Soil and groundwater are the media of concern. Contaminated soil and sediment was excavated, treated, and disposed of off site. No active groundwater remediation is currently required; natural attenuation will augment the on-site dewatering pumps. Energy Research will be the site steward starting in FY 2000.
Site A/Plot M	Site A was returned to the Forest Preserve District of Cook County, IL in FY 1997 for unimproved recreational use by the public. Plot M, which was capped in 1973, was returned to the Forest Preserve in 1956 with ongoing surveillance and monitoring (S&M) performed by DOE. S&M activities are being transferred to the DOE Grand Junction Office by FY 1998.

Junction Office by the end of Fiscal Year 1998. Also, the Chicago Operations Office is responsible for payments to support the Princeton Site A/B Project. The responsibility for the payments will be transferred to the Office of Energy Research prior to FY 2006.

E.3.2 Cost and Completion Dates

The Chicago Operations Office has divided its environmental management work into 20 discrete projects. A Project Baseline Summary (PBS) exists for each project and contains detailed programmatic information, including cost, schedule, end state, and interim milestones. A summary of the Chicago cost and schedule information is illustrated in Exhibit E-11. For additional information about these projects, refer to individual PBSs.

The estimated EM life-cycle cost of the Chicago Operations Office site cleanups is $0.3 billion (constant 1998 dollars). This estimate does not include approximately $1.1 billion (constant 1998 dollars) of non-EM costs. Overall site completion dates for EM work scope are as follows:

Site	Date
Ames Laboratory	1999
Argonne National Laboratory - East	2002
Argonne National Laboratory - West	2000
Brookhaven National Laboratory	2006
Fermi National Accelerator Laboratory	1997
Princeton Plasma Physics Laboratory	1999
Site A/Plot M	1997

Exhibit E-11 Chicago Operations Office
Cleanup Project Summary: Duration and Costs (All costs in thousands of 1998 dollars)

Site Closure Project Activities	1997 - 2006	2007 - 2070	Total	97	98	99	00	01	02	03	04	05	06	07	08	09	10	11	12	13	14	15	16	17	18	19	20	21	22	23	24
Chicago Operations Office																															
Site A Cleanup	350	0	350																												
Surveillance and Maintenance Activities	490	0	490																												
Princeton Site A/B Payments	2,445	0	2,445																												
Chicago Operations Program Support	3,859	0	3,859																												
Ames Laboratory																															
Ames Remedial Actions	237	0	237																												
Ames Waste Operations	785	0	785																												
Argonne National Laboratory-East (ANL-E)																															
ANL-E Waste Operations	25,315	0	25,315																												
ANL-E Remedial Actions	21,605	0	21,605																												
ANL-E Decontamination & Decommissioning Actions	30,390	0	30,390																												
ANL-E Program Management	3,980	0	3,980																												
Argonne National Laboratory-West (ANL-W)																															
ANL-W Waste Operations	6,838	0	6,838																												
ANL-W Remedial Actions	6,856	0	6,856																												

Exhibit E-11 Chicago Operations Office (Continued)
Cleanup Project Summary: Duration and Costs (All costs in thousands of 1998 dollars)

Site Closure Project Activities	1997 - 2006	2007 - 2070	Total
Brookhaven National Laboratory (BNL)			
BNL Waste Operations	17,250	0	17,250
BNL Boneyard	8,134	0	8,134
BNL Decontamination and Decommisioning Actions	31,189	0	31,189
BNL Remedial Actions	130,723	0	130,723
BNL Program Management	22,751	0	22,751
Fermi National Accelerator Laboratory (FNAL)			
FNAL Waste Operations	2,157	0	2,157
Princeton Plasma Physics Laboratory (PPPL)			
PPPL Waste Operations	9,017	0	9,017
PPPL Remedial Actions	1,535	0	1,535
Total	**325,906**	**0**	**325,906**

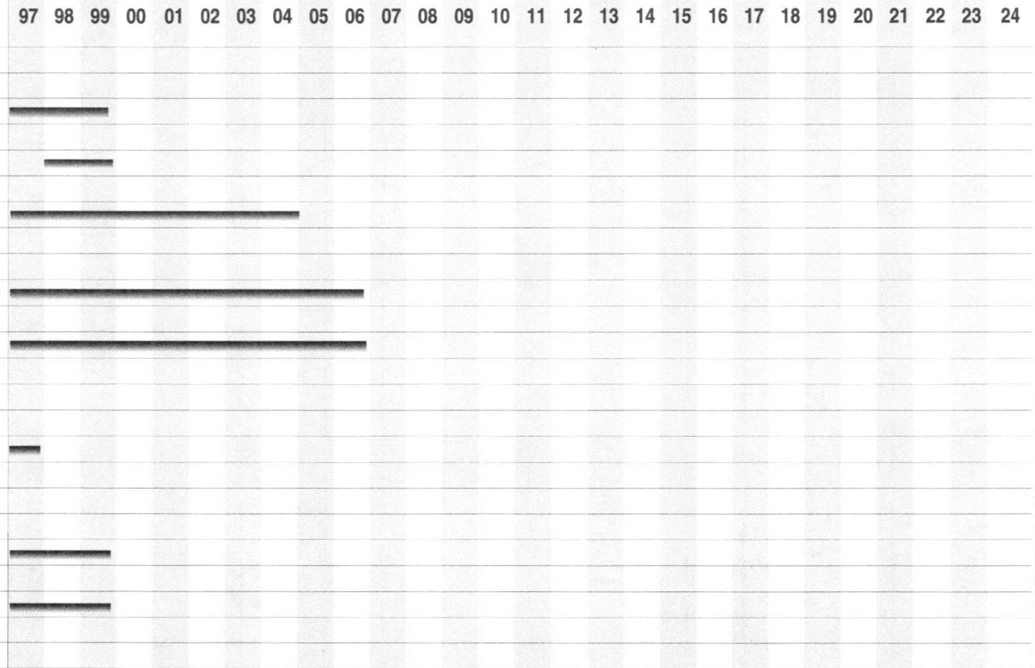

The projected cost profile for environmental management associated with the Chicago Operations Office is developed by combining the cost estimates in each of the PBSs. Exhibit E-12 displays the resultant baseline cost profile.

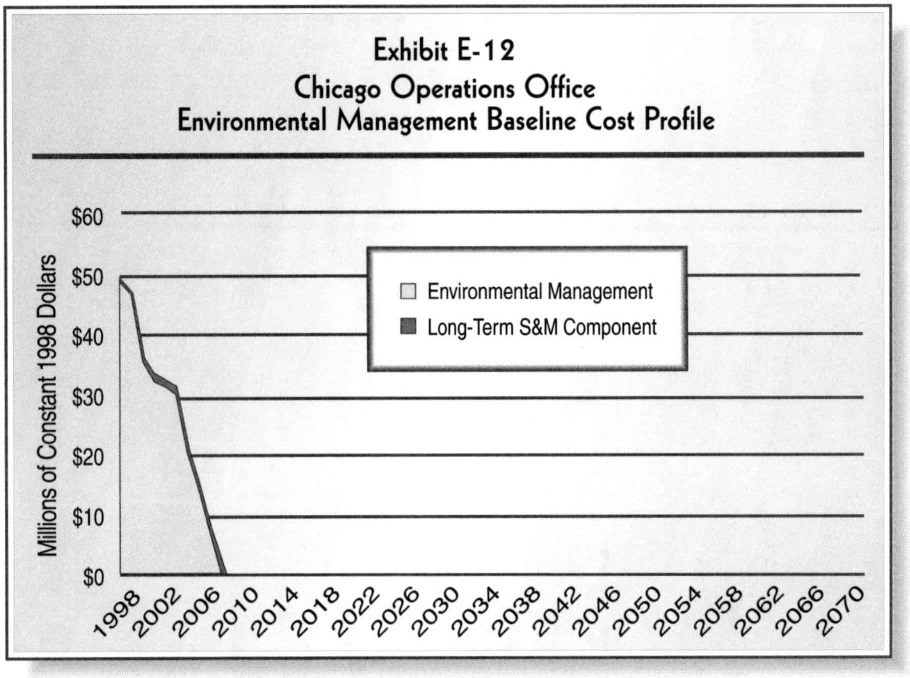

E.3.3 Work Scope Summary

Cleanup activities at the sites managed by the Chicago Operations Office include the management of groundwater contaminated with radionuclides and hazardous substances and soils and debris contaminated with radionuclides at Brookhaven National Laboratory, and rubble & debris contaminated with hazardous substances at Argonne National Laboratory-East. The sections below describe the major waste, material, and contaminated media volumes to be addressed by the Chicago Operations Office. The volumes reported are approximate, and correspond to the major waste, material, and media flows, potential treatment processes, and off-site disposal destinations presented in Exhibit E-13, the Chicago Operations Office Conceptual Summary Disposition Map.

Transuranic Waste

- Approximately 80 cubic meters of transuranic waste are currently in inventory and 5.1 cubic meters are expected to be generated over the life cycle of operations. After treatment and repackaging, 82 cubic meters are expected to be disposed of at the Waste Isolation Pilot Plant (WIPP). A disposition path has not been determined for 2.5 cubic meters of transuranic waste.

Exhibit E-13

Chicago Operations Office Conceptual Summary Disposition Map

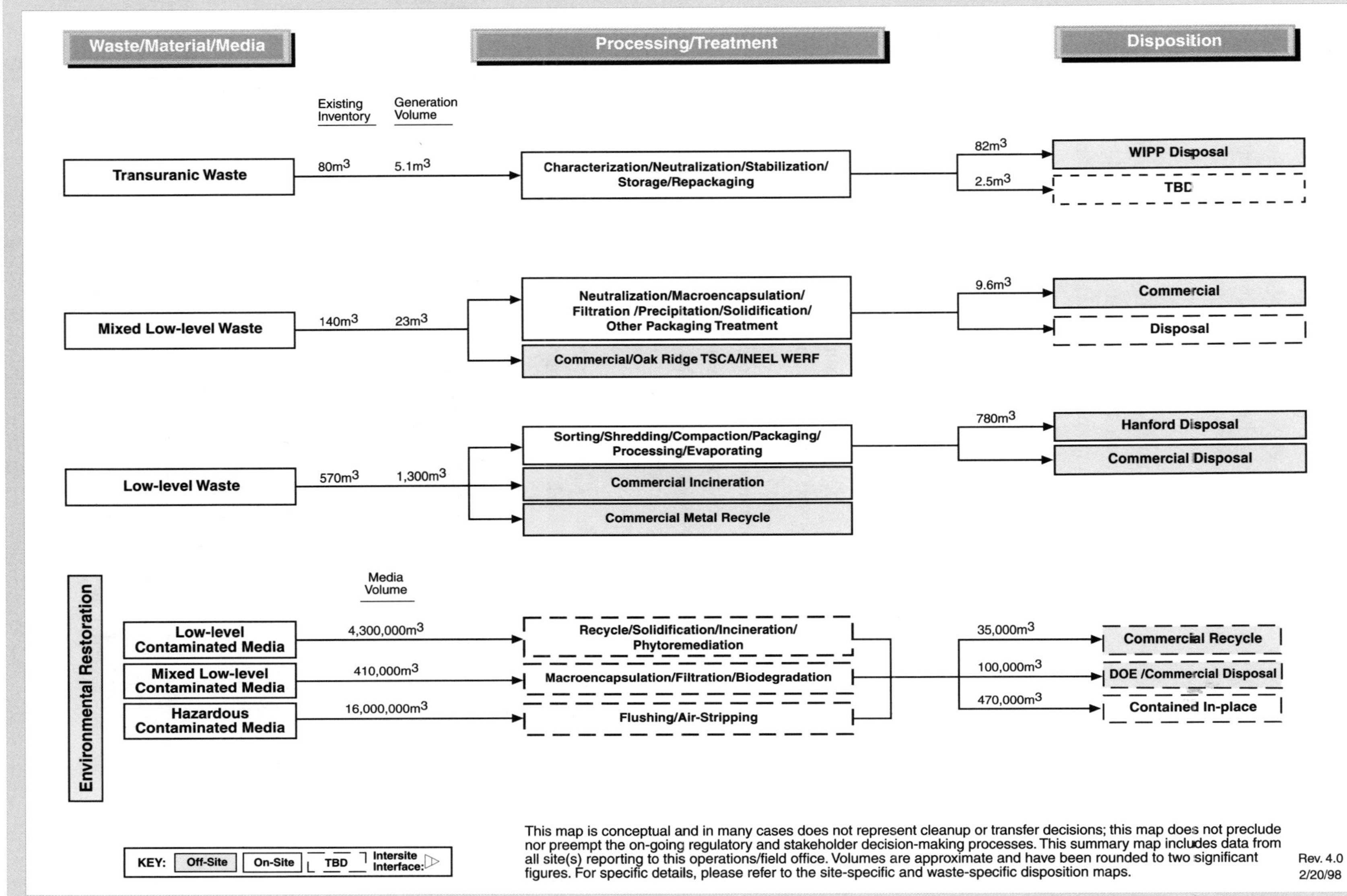

This map is conceptual and in many cases does not represent cleanup or transfer decisions; this map does not preclude nor preempt the on-going regulatory and stakeholder decision-making processes. This summary map includes data from all site(s) reporting to this operations/field office. Volumes are approximate and have been rounded to two significant figures. For specific details, please refer to the site-specific and waste-specific disposition maps.

Rev. 4.0
2/20/98

Other Waste

- Approximately 140 cubic meters of mixed low-level waste are currently in inventory and 23 cubic meters are expected to be generated annually. Waste will undergo a range of treatment activities as well as incineration at other DOE sites. After treatment, 9.6 cubic meters are expected to be disposed of at an off-site commercial facility.
- Nearly 570 cubic meters of low-level waste are currently in inventory and 1,300 cubic meters of low-level waste are expected to be generated annually. Waste will undergo a range of treatment activities as well as incineration and recycling at off-site commercial facilities. After treatment, 780 cubic meters are expected to be disposed of at Hanford, and additional volumes are expected to be disposed of at an off-site commercial facility.

Remedial Action and Facility D&D

- A total of 20.7 million cubic meters of contaminated environmental media will be managed through a variety of remedial responses. This volume includes 4.3 million cubic meters of soils, rubble & debris contaminated with radionuclides, 410,000 cubic meters of soils and sediments contaminated with radionuclides and hazardous substances, and 16 million cubic meters of groundwater contaminated with hazardous substances. After a range of treatment activities, 35,000 cubic meters are expected to be sent to an off-site commercial recycling facility, 100,000 cubic meters are expected to be disposed of at an off-site DOE facility and an off-site commercial facility, and 470,000 cubic meters are expected to be contained in place.

Exhibit E-14 illustrates the Chicago Operations Office environmental management costs by major work scope categories.

Exhibit E-14
Chicago Operations Office
Life-Cycle Costs by Category

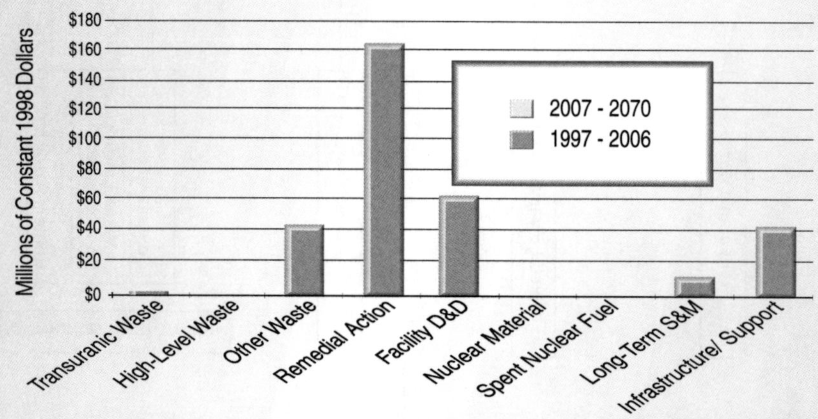

E.3.4 Critical Closure Path and Programmatic Risk

The critical closure path schedule, presented as Exhibit E-15, sets forth the timetable for completing the closure activities at the Chicago Operations Office. In the exhibit, the bars represent critical activities, and the diamonds represent critical events/milestones. The critical closure path identifies the major cleanup activities that have little scheduling flexibility and must occur without delays if the EM cleanup mission is to be completed by 2006.

Completion of the EM mission at the Chicago Operations Office as scheduled will depend on the timely accomplishment of critical activities and events. Sites have assigned programmatic risk scores to each of the critical activities/milestones. Appendix D provides a complete definition of programmatic risk. Exhibit E-16 presents a summary of activities/milestones on the critical closure path that have high programmatic risk (programmatic risk scores of 4 or 5 in any category).

Exhibit E-15 **Chicago Operations Office Critical Closure Path** (Timelines and Risk Data)

Activity Description	Project	97–08 Timeline (Risk)
Argonne National Laboratory - East (ANL-E)		
Juggernaut, ZPR/ATSR, Building 310 and Building 301 Projects	ANL-E Decontamination & Decommissioning Actions	(1,3,4)
Program Management is a Level-of-effort Planning Activity	ANL-E Program Management	(1,1,1)
Submission of No Further Action Requests for all Release Sites	ANL-E Remedial Actions	(1,3,2)
Argonne National Laboratory - West (ANL-W)		
ANL-W Remedial Actions	ANL-W Remedial Actions	(2,3,3)
Brookhaven National Laboratory (BNL)		
Complete Shipment	BNL Boneyard Waste	Complete Shipment (2,4,4)
OUVI - EDB	BNL Remedial Actions	(1,1,1)
Public Water Hookups	BNL Remedial Actions	(1,1,1)
RA6 - Landfills Phase III	BNL Remedial Actions	(1,1,1)
RA3 - Cesspools	BNL Remedial Actions	(1,1,1)
OUII - BGRR, AGS, Scrapyard	BNL Remedial Actions	(1,1,1)
OUI - HWMF	BNL Remedial Actions	(2,4,5)
OUIV - Central Steam Facility	BNL Remedial Actions	(1,1,1)
RA5 - Groundwater Removal	BNL Remedial Actions	(1,1,1)
OUV - Sewage Treatment Plant	BNL Remedial Actions	(2,5,5)
Projectwide Support	BNL Program Management	(1,1,1)
OUIII - Accelerated Action Offsite	BNL Remedial Actions	(1,1,1)
Chicago Operations Office		
Turn responsibility for Site A/Plot M S&M over to Grand Junction Project Office	Surveillance and Maintenance Activities	(1,1,1)
Turn responsibility for PRP payments over to ER	Princeton Site A/B Payments	(1,3,1)

Event/Milestone — Critical Activity

Programmatic Risk Categories → (3,2,3) (Technological, Work Scope Definition, Inter-Site Dependency) Range = 1 to 5 where 5 equals highest risk

Exhibit E-16

Summary of High Programmatic Risk Activities/Milestones: Chicago Operations Office

Site	Project, Activity, Event	Start/End Dates	Programmatic Risk Categories		
			Technological	Work Scope Definition	Intersite Dependency
Argonne National Laboratory - East	Juggernaut, ZPR/ATSR, Bldg. 310 and Bldg. 301 projects	Aug 99/ Sep 02	1	3	4
Brookhaven National Laboratory	OU-III - Source Areas	Oct 96	2	4	2
	Complete Shipments	Aug 00	2	4	4
	OU-I - HWMF	Oct 96/ Sep 00	2	4	5
	OU-V - Sewage Treatment Plant	Oct 96/ Aug 06	2	5	5

E.4 Idaho Operations Office Summary

The Idaho Operations Office manages environmental management activities at the Idaho National Engineering and Environmental Laboratory (INEEL), a site that occupies 890 square miles in a remote desert area in southeastern Idaho. The Laboratory consists of 9 major operating areas at the site and several facilities in the City of Idaho Falls, located 42 miles east of the Laboratory.

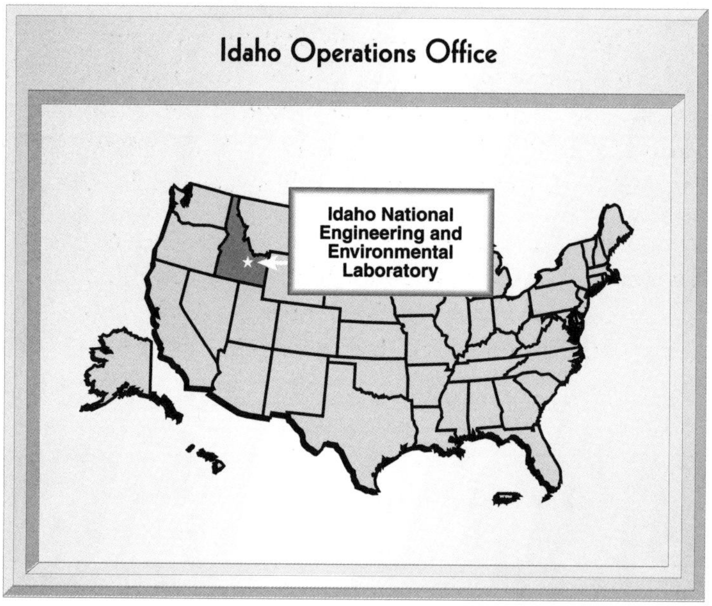

INEEL is committed to completing several Comprehensive Environmental Response, Compensation, and Liability Act (CERCLA) remediation sites by FY 2006, while pursuing longer-term projects to accomplish cleanup of transuranic and high-level wastes, spent nuclear fuel disposition, and closure of remaining CERCLA remediation sites after FY 2006.

In addition to completing the environmental management mission in Idaho, INEEL has implemented a long-range plan which will transform the laboratory from a Department of Energy (DOE) multi-program national laboratory focused on site cleanup to a national multi-program engineering and environmental laboratory. The near-term focus of the long-range plan is to support key capabilities and competitiveness necessary to ensure INEEL's future by leveraging the cleanup mission and making other long-term investments.

E.4.1 End State

INEEL's final end state is described in the INEL Comprehensive Facilities and Land Use Plan issued March 1996. The laboratory will continually work with their stakeholders and jointly review the Land Use Plan for accuracy and adequacy. The end state objective at INEEL is to complete cleanup per Federal Facility Agreement and Consent Order requirements and disposition all waste and other materials in accordance with existing and future agreements; meet the milestones of the Idaho Settlement Agreement; and complete the work covered under the Site Treatment Plan.

INEEL is planning to restore its site to an industrial and open space end state based on an analysis of site land use for the next 100 years. The site will contain an on-site disposal cell for contact-handled low-level waste. Currently, the site is also planning to store spent nuclear fuel until 2035, and treat and store high-level waste until 2070. High-level waste will be ready for shipment in 2035.

E.4.2 Cost and Completion Profile

Idaho Operations Office has divided its environmental management work into 43 discrete projects. A Project Baseline Summary (PBS) exists for each project and contains detailed programmatic information, including cost, schedule, scope, end state, and interim milestones. A summary of the cost and schedule information for these projects is illustrated in Exhibit E-17.

The estimated life-cycle cost of DOE's Environmental Management (EM) program's cleanup mission for Idaho is $16.3 billion (constant 1998 dollars) with the last project ending in September 2070. However, the majority of the work scope will be completed by 2050, with only monitoring and other essential functions continuing beyond 2050.

Exhibit E-17 Idaho Operations Office
Cleanup Project Summary: Duration and Costs (All costs in thousands of 1998 dollars)

Site Closure Project Activities	1997 - 2006	2007 - 2070	Total
Idaho Operations Office			
Science and Technology Coordination	15,778	0	15,778
Idaho National Engineering and Environmental Laboratory (INEEL)			
INEEL Medical Facilities	270	0	270
INEEL Emergency Response Facilities	768	0	768
Security Facilities Consolidation Project	6,930	0	6,930
INEEL Electrical Distribution Upgrade	10,152	0	10,152
INEEL Road Rehabilitation	11,047	0	11,047
Constructed New Facilities	30,235	0	30,235
Pit 9 Remediation	133,957	0	133,957
Central Facilities Area (CFA) Remediation	9,969	9,208	19,176
Power Burst Facility/Auxiliary Reactor Area	12,978	2,498	15,476
Electrical and Utility Systems Upgrade (EUSU) Project, ICPP	56,672	0	56,672
AMWTP Asset Acquisition Project (Privatized)	517,100	52,300	569,400
Health Physics Instrument Laboratory	11,703	0	11,703
INEEL LLW/MLLW/Other Waste Program	193,490	0	193,490

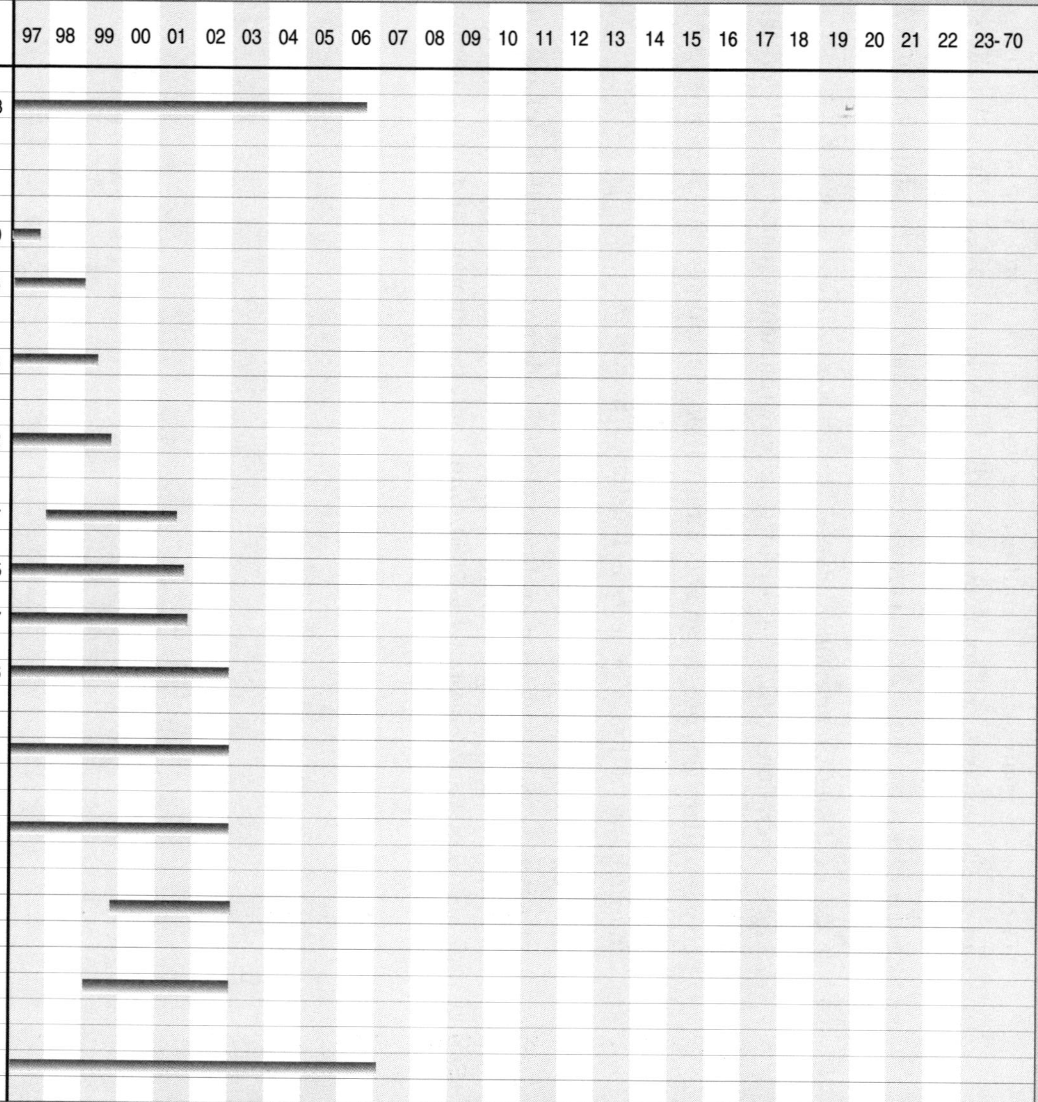

Exhibit E-17 Idaho Operations Office (Continued)
Cleanup Project Summary: Duration and Costs (All costs in thousands of 1998 dollars)

Site Closure Project Activities	1997 - 2006	2007 - 2070	Total	97 98 99 00 01 02 03 04 05 06 07 08 09 10 11 12 13 14 15 16 17 18 19 20 21 22 23-70
Dry Transfer and Storage Project (Privatized)	83,139	0	83,139	
National LLW Program	36,087	0	36,087	
INEEL Transuranic Waste	281,346	0	281,346	
Integrated Waste Operations Program	111,171	0	111,171	
Pre-FY 2007 Surplus Facility Deactivation Project	108,249	0	108,249	
LLW/MLLW Center of Excellence	16,604	0	16,604	
Pre-2007 INEEL Surveillance and Maintenance (S&M)	35,923	0	35,923	
Sitewide Monitoring Area Remediation	52,741	12,014	64,755	
High-Level Waste Pretreatment	478,618	376,645	855,263	
National Spent Nuclear Fuel (SNF) Program	157,479	71,697	229,176	
AMWTP Production Operations	124,815	208,558	333,373	
Rad Waste Management Complex Remediation	441,933	905,815	1,347,748	Planned Completion Date is 2024
Low Activity Waste Treatment	61,818	116,579	178,396	Planned Completion Date is 2025
Test Area North Remediation	79,452	15,430	94,882	Planned Completion Date is 2026
Integrated SNF Program	159,251	179,045	338,297	Planned Completion Date is 2034
Emptied SNF Facilities	295,596	400,202	695,798	Planned Completion Date is 2035

Exhibit E-17 Idaho Operations Office (Continued)
Cleanup Project Summary: Duration and Costs (All costs in thousands of 1998 dollars)

Site Closure Project Activities	1997 - 2006	2007 - 2070	Total
HLW Immobilization Facility	0	1,533,249	1,533,249
HLW Treatment and Storage	173,852	2,190,700	2,364,552
Test Reactor Area Remediation	7,717	10,487	18,204
Decontamination and Decommissioning	127,011	339,079	466,090
Post-FY2006 Surplus Facility Deactivation Projects	0	31,757	31,757
INEEL Site-Wide Environmental Protection	60,804	316,158	376,962
Long-Term Treatment/ Storage/Disposal Operations	0	937,878	937,878
Post-2006 Surveillance, Maintenance, and Monitoring	0	52,260	52,260
Vitrified HLW Storage	0	21,499	21,499
Site Wide Landlord Operations	394,181	1,433,618	1,827,799
Remediation Operations	161,913	225,844	387,757
Idaho Chemical Processing Plant Non-Process Plant Operations	488,807	1,639,419	2,128,226
Idaho Chemical Processing Plant Remediation	114,783	198,736	313,519
Total	**5,064,340**	**11,280,676**	**16,345,016**

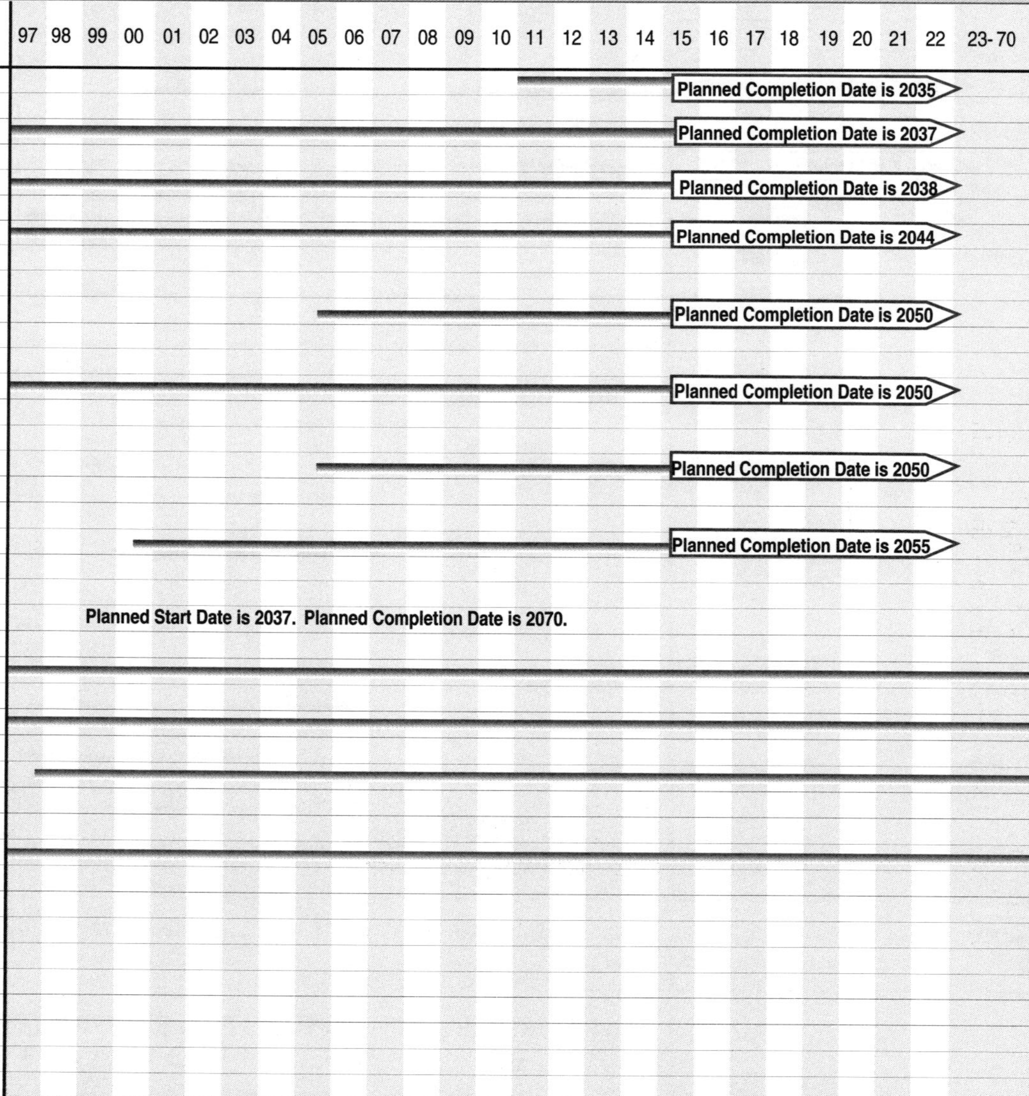

The projected cost profile for environmental management associated with the Idaho Operations Office is developed by combining the cost estimates in each of the PBSs. Exhibit E-18 displays the resultant baseline cost profile.

Exhibit E-18
Idaho Operations Office Environmental Management Baseline Cost Profile

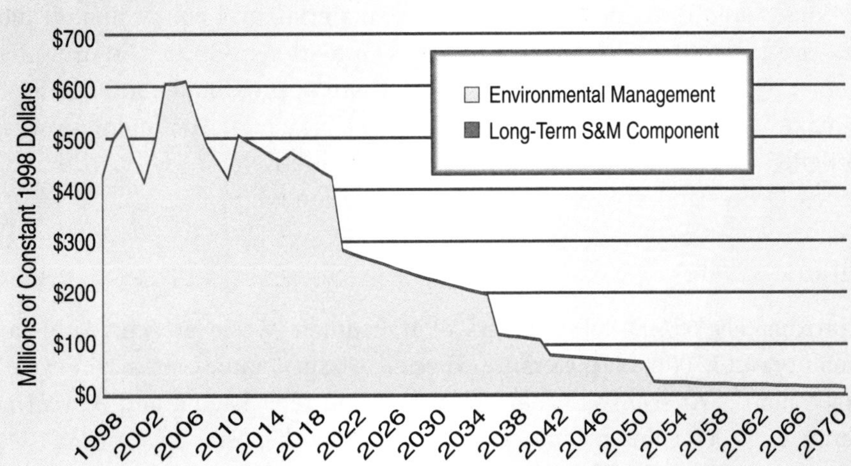

E.4.3 Work Scope Summary

The Idaho cleanup mission requires projects to accomplish the cleanup of transuranic and high-level wastes, the disposition of spent nuclear fuel, and the cleanup and closure of CERCLA remediation sites.

Work is conducted using the seven criteria established by the EM program: (1) eliminate the most urgent risks; (2) reduce "mortgage" and support costs to free up funds for further risk reduction; (3) protect worker health and safety; (4) reduce the generation of wastes; (5) create a collaborative relationship between DOE, its regulators, and its stakeholders; (6) focus science and technology development on cost and risk reduction; and (7) integrate spent nuclear fuel and waste treatment and disposal across INEEL. The Laboratory has four programs in place to complete its environmental mission:

1. The Waste Management program will treat, store, and dispose of low-level waste, mixed low-level waste, transuranic waste, and high-level waste in compliance with agreements and the Site Treatment plan.

2. The Environmental Restoration program will remediate all Federal Facility Agreement/Consent Order (FFA/CO) identified contaminated land/facilities as determined under CERCLA. Contaminated facilities used for previous

INEEL nuclear reactor testing, spent nuclear fuel reprocessing, and waste treatment, storage, and disposal will undergo decontamination and decommissioning (D&D).

3. The Nuclear Materials and Facilities Stabilization program will receive and store spent nuclear fuel until final disposition.

4. The Infrastructure and Deactivation programs ensure adequate infrastructure support for the above-mentioned programs.

The sections below describe the major waste, material, and contaminated media volumes to be addressed by the Idaho Operations Office Environmental Management program. The volumes reported are approximate, and correspond to the major waste, material, and media flows, potential treatment processes, and off-site disposal destinations presented in Exhibit E-19, the Idaho Operations Office Conceptual Summary Disposition Map.

Transuranic Waste

- Approximately 65,000 cubic meters of transuranic waste are currently in inventory and 3,700 cubic meters are expected to be generated over the life cycle of operations. After on-site characterization and repackaging and AMWTP treatment, 50,000 cubic meters are expected to be disposed of at the Waste Isolation Pilot Plant (WIPP).

High-level Waste

- Approximately 35 cubic meters of HEPA filters are expected to be received from ANL-W. Currently, there are 10,000 cubic meters in inventory. Nearly 11,000 cubic meters of high-level waste are expected to be generated over the life cycle of operations.
- After removal of high-level waste, 11 tanks and 42 bins are expected to be stabilized and closed.

Other Waste

- Approximately 22,000 cubic meters of low-level waste are expected to be received from off site. Currently, there are 9,400 cubic meters of low-level waste in inventory. Over 100,000 cubic meters of low-level waste are expected to be generated over the life cycle of operations. After treatment, including on-site and commercial stabilization, compaction, and incineration, the low- level waste is expected to be disposed of at an undetermined off-site low-level waste disposal facility and at the on-site Radioactive Waste Management Complex (RWMC) disposal facility.

Exhibit E-19 Idaho Operations Office Conceptual Summary Disposition Map

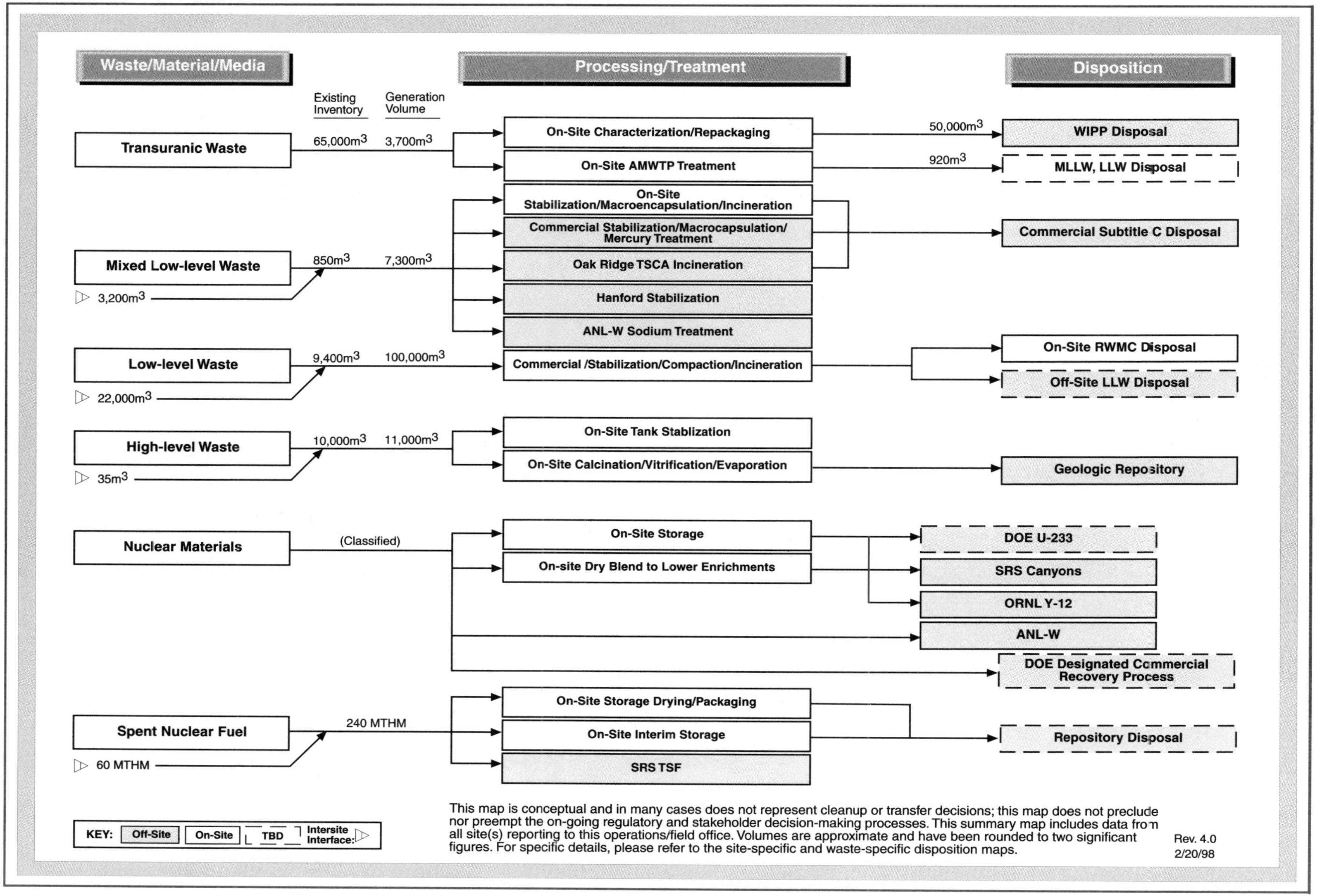

Exhibit E-19 Idaho Operations Office Conceptual Summary Disposition Map (Continued)

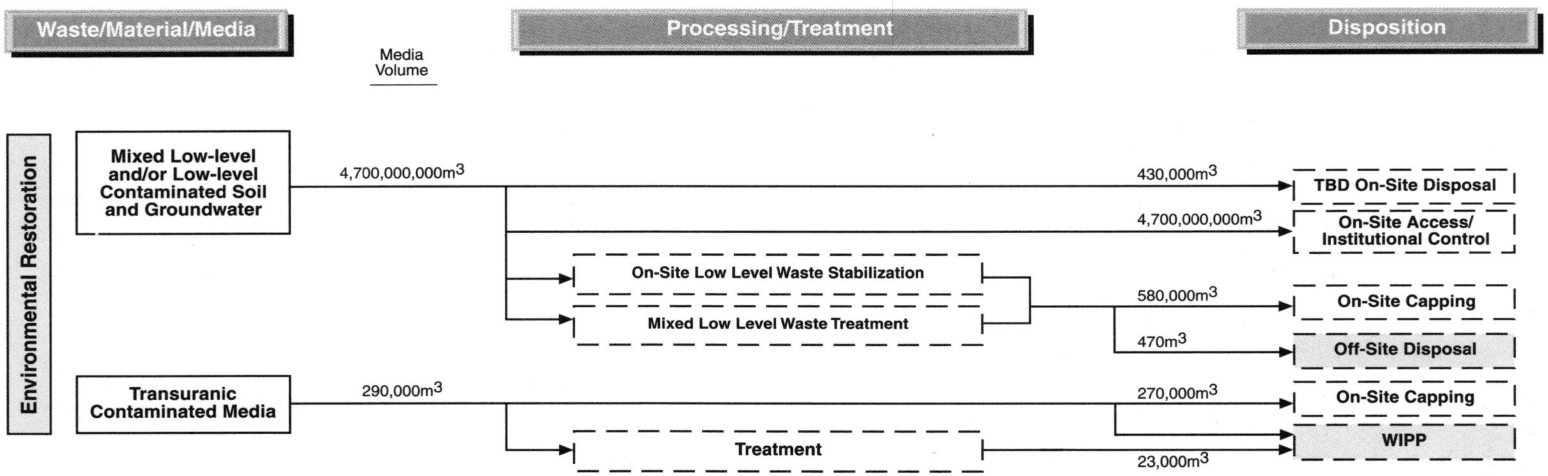

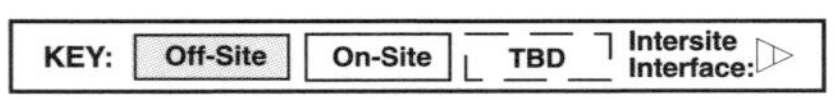

This map is conceptual and in many cases does not represent cleanup or transfer decisions; this map does not preclude nor preempt the on-going regulatory and stakeholder decision-making processes. This summary map includes data from all site(s) reporting to this operations/field office. Volumes are approximate and have been rounded to two significant figures. For specific details, please refer to the site-specific and waste-specific disposition maps.

Rev. 4.0
2/20/98

- Approximately 3,200 cubic meters of mixed low-level will be received from off site. Currently, there are 850 cubic meters of mixed low-level waste in inventory. Approximately 7,300 cubic meters of mixed low-level waste are expected to be generated over the life cycle of operations. After treatment, an undetermined amount of treatment residues are expected to be disposed of at an off-site commercial Subtitle C disposal facility.

Remedial Action and Facility D&D

- Approximately 4.7 billion cubic meters of mixed low-level and low-level contaminated environmental media will be managed through a variety of remedial response strategies: following stabilization and treatment, 580,000 cubic meters are expected to be capped on site and 470 cubic meters are expected to be disposed of off site; 430,000 cubic meters are expected to be disposed of at an undetermined on-site disposal facility, and 4.7 billion cubic meters will remain on site under access/institutional controls.

- Approximately 290,000 cubic meters of environmental media contaminated with transuranic elements will be processed. After treatment, 270,000 cubic meters are expected to be capped in- place and 23,000 cubic meters are expected to be disposed of at WIPP.

Nuclear Material

- Nuclear materials quantities are classified and cannot be disclosed in this document.

Spent Nuclear Fuel

- Approximately 60 metric tons heavy metal of spent nuclear fuel will be received from off-site sources. Currently, there are 240 cubic meters of spent nuclear fuel in inventory. After on-site storage, drying, and packaging, an undetermined quantity of spent nuclear fuel is expected to be shipped off site to a repository for disposal.

Exhibit E-20 shows the distribution of life-cycle costs by major work scope category for the Idaho Operations Office.

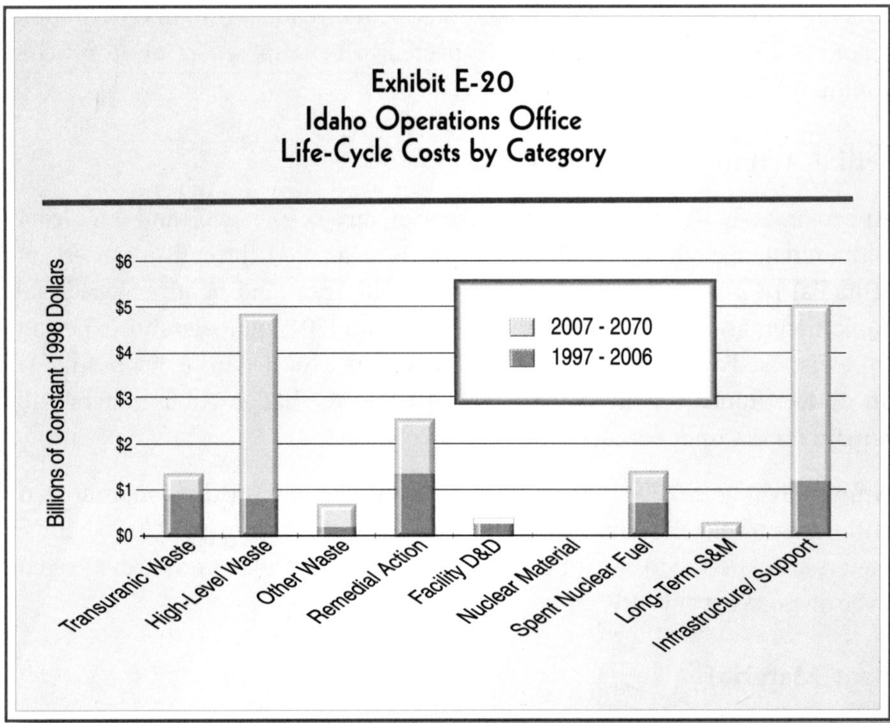

E.4.4 Critical Closure Path and Programmatic Risk

The critical closure path schedule presented as Exhibit E-21 sets forth the timetable for completing the closure activities at the Idaho Operations Office. The highlighted activities show the critical closure path, which represents the series of events that drive the overall completion date for the site and must occur without delay if the EM cleanup mission at INEEL is to meet the requirements of the Idaho Settlement Agreement, other regulatory compliance agreements, and court orders. In Exhibit E-21, the bars represent critical activities, and the triangles represent critical events/milestones.

Completion of the EM mission at the Idaho Operations Office as scheduled will depend on the timely accomplishment of critical activities and events. Sites have assigned programmatic risk scores to each of the critical activities/milestones. Appendix D provides a complete definition of programmatic risk. Exhibit E-22 presents a summary of activities/milestones on the critical closure path that have high programmatic risk (programmatic risk scores of 4 or 5 in any category). The Idaho Operations Office version of *Paths to Closure* provides more details on the management approach for these high programmatic risk issues.

Exhibit E-21

Idaho Operations Office Critical Closure Path

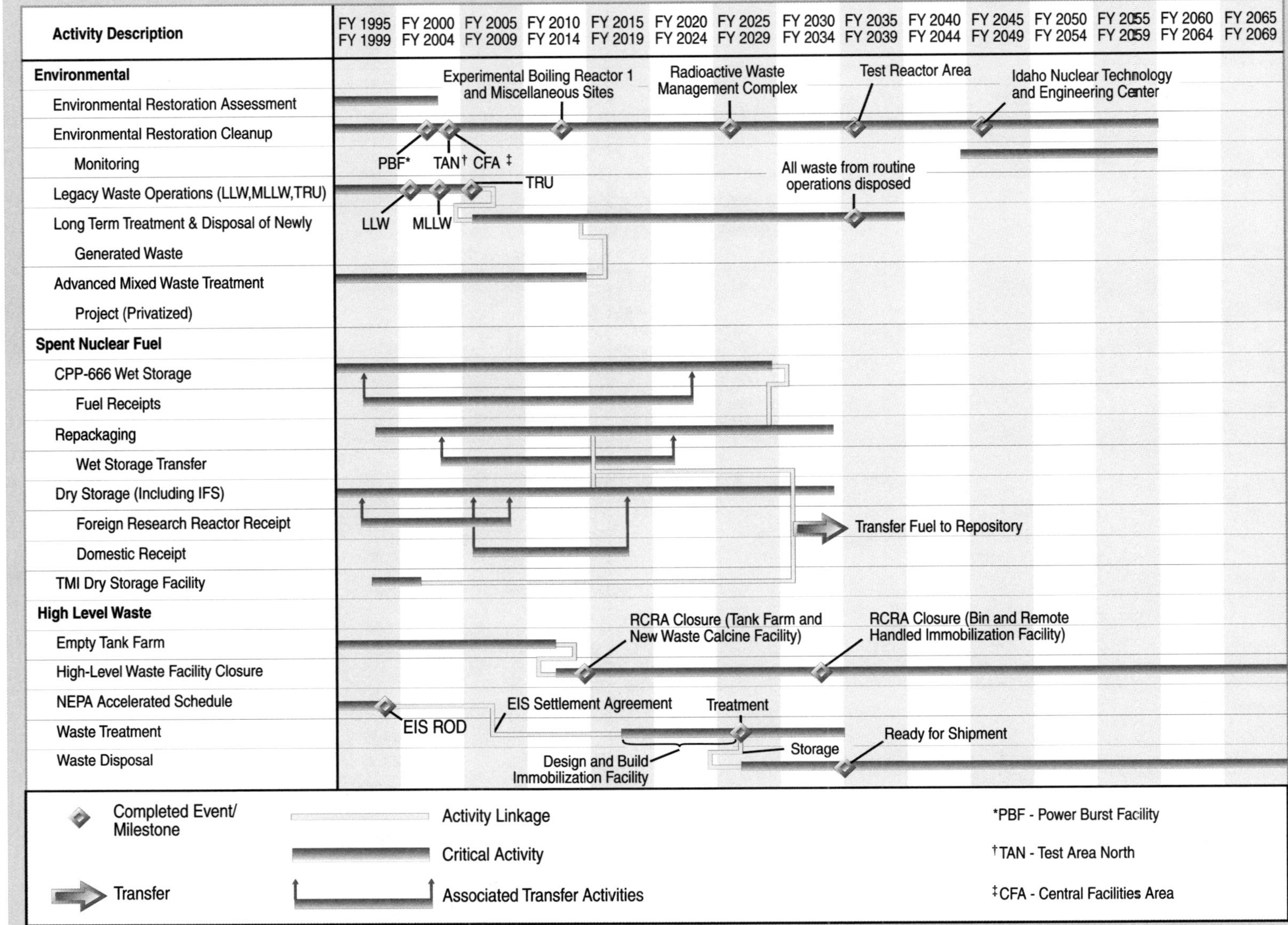

Exhibit E-21

Idaho Operations Office Critical Closure Path (Continued)

Milestone Description	FY 1995 FY 1999	FY 2000 FY 2004	FY 2005 FY 2009	FY 2010 FY 2014	FY 2015 FY 2019	FY 2020 FY 2024	FY 2025 FY 2029	FY 2030 FY 2034	FY 2035 FY 2039	FY 2040 FY 2044	FY 2045 FY 2049	FY 2050 FY 2054	FY 2055 FY 2059	FY 2060 FY 2064	FY 2065 FY 2069
First WIPP Data Complete for TRU Waste	3/98														
WIPP Accepts TRU Waste	5/98														
EPA Authority Granted for WIPP	6/98														
NRC License Issued (TMI)	10/98														
RW Record of Decision Issued for Repository		9/00													
NRC Approval of Safety Analysis Report Amendment 18		1/01													
RW Submits License Application to NRC for Repository		2/02													
Data Package Preparation Guideline for SNF Disposal Issued to the INEEL				12/09											
Repository Open for Commercial SNF License to Operate Approved				1/10											
RW Issues Final Disposabilty Interface Specifications (DIS)				1/10											
Repository Open for DOE SNF					1/15										
Reissue Guidance Document (First SNF Shipments Information)					1/15										
National Repository Ready to Receive DOE Fuel					1/15										

◆ Event/ Milestone

Exhibit E-22
Summary of High Programmatic Risk Activities/Milestones: Idaho Operations Office

Project, Activity, Event	Start/End Dates	Programmatic Risk Categories		
		Technological	Work Scope Definition	Intersite Dependency
Issue a Record of Decision for shipment and ultimate disposal of SNF outside Idaho	Apr 99	2	3	5
First TRU Waste Shipment to WIPP	Apr 99	1	1	5
Convert pretreated waste to final disposable form	Oct 20/ Sep 35	4	3	4
Store vitrified waste containers until repository is ready	Oct 20/ Sep 70	2	3	4

E.5 Nevada Operations Office Summary

For over 40 years, the primary mission of the Department of Energy's Nevada Operations Office (DOE/NV) was to conduct research, development, and testing of nuclear devices. Most testing took place at the Nevada Test Site, but nuclear testing activities have also been conducted at eight off-site locations in five different states.

Nevada Operations Office

The Nevada Operations Office manages facilities on the Nevada Test Site, which covers 1,350 square miles of land about 65 miles northwest of Las Vegas, Nevada. In addition, the Nevada Operations Office is responsible for environmental management at off-site test areas where radiological contamination has occurred including: Amchitka Island (Alaska); Central Nevada Test Site (Nevada); Project Chariot (Alaska); Project Gasbuggy (New Mexico); Project Gnome-Coach (New Mexico); Project Rulison (Colorado); Project Rio Blanco (Colorado); Project Salmon (Mississippi); and Project Shoal (Nevada).

The ***Nevada Test Site (NTS)*** is located in a remote region of Nevada and is roughly the size of the State of Rhode Island. In addition to weapons testing, the Nevada Test Site has hosted secondary missions including: neutron and gamma-ray interaction studies; open air reactor, nuclear engine, and nuclear furnace tests; hazardous materials spill response testing; and a variety of other experiments involving radioactive and non-radioactive materials conducted by the Department of Defense.

Amchitka Island was the site of three underground nuclear detonations conducted in October 1965, October 1969, and November 1971. These tests were conducted for seismic testing, calibration, and warhead development.

The ***Central Nevada Test Site*** was used for one subsurface nuclear test, Project Faultless, detonated in January 1968. The Department conducted the test to determine the suitability of the area for additional testing. It also conducted nonnuclear special experiments to determine the behavior of seismic waves.

The ***Gnome-Coach*** and ***Gasbuggy Sites*** were part of the Plowshare program, which was a series of nuclear and conventional tests conducted by the Atomic Energy Commission to explore peacetime uses of nuclear explosives. The Project Gnome test was conducted in bedded salt in December 1961. The Gasbuggy Site was the location of a single subsurface nuclear test in December 1967.

The ***Rio Blanco*** and ***Rulison*** tests, also part of the Plowshare program, were designed to increase natural gas production from low-permeability sandstone. The Project Rulison detonation took place in September 1969 in a sandstone formation. The Project Rio Blanco consisted of the nearly simultaneous detonation of three devices in a deep well in May 1973.

The ***Salmon Site*** was used for two nuclear detonations, Salmon and Sterling, to evaluate the seismic response of salt deposits to nuclear explosives. Salmon Site was also the location for two nonnuclear gas detonations used for seismic decoupling studies in the Miracle Play Program. The Department conducted the Salmon test in the Tatum Salt Dome in October 1964. It detonated the Sterling test in the Salmon cavity in December 1966.

The ***Project Shoal Site*** nuclear test was conducted in October 1963. The purpose of the test was to determine the effect of a nuclear detonation in a granite rock formation and to compare the seismic activity of natural earthquakes with activity from an underground nuclear explosion.

The ***Tonopah Test Range***, northwest of the Nevada Test Site, is used by the Department of Energy's Albuquerque Operations Office and the Department of Defense for research and development of ordnance delivery systems, electronic combat training missions, and other activities. The Nevada Operations Office has environmental restoration responsibilities for historic DOE/NV testing activities conducted at the site. For planning and control purposes, the Tonopah Test Range is considered to be part of the NTS.

E.5.1 End State

The Nevada Test Site is a Defense Programs site. The primary mission of the site is nuclear stockpile stewardship including the maintenance of readiness to conduct underground nuclear tests as directed. Decisions regarding future land use on the Nevada Test Site are awaiting completion of the Resource Management Plan, which is scheduled for completion in October 1998. Future land uses for the Nevada Test Site, as well as potential uses of facilities that are to be decontaminated and decommissioned are being developed at this time in compliance with commitments contained in the Nevada Test Site Environmental Impact Statement. Decisions involving resource management, future land use, and private development will be done in partnership with the interests of the Department of Energy, national laboratories, the U.S. Air Force, the Bureau of Land Management, Tribal Nations, State and local agencies, and stakeholders.

Responsibility for land use on the Tonapah Test Range falls within the purview of the Department of Defense, U.S. Air Force. The Department of Defense is in the process of developing an Environmental Impact Statement governing Air Force activities on the Nellis Air Force Range, which includes the Tonapah Test Range.

The Off-sites Projects

Amchitka Island (Alaska), Project Rio Blanco (Colorado), Project Rulison (Colorado), Project Salmon (Mississippi), Central Nevada Test Site (Nevada), Project Shoal (Nevada), Project Gasbuggy (New Mexico), and Project Gnome-Coach (New Mexico) will have surface areas released for alternative uses without restriction and/or relinquished to the U.S. Fish and Wildlife Service (Amchitka), the State of Mississippi (Project Salmon), or the U.S. Bureau of Land Management. The subsurface will require controlled access. Environmental monitoring of the surface areas, if necessary, will be implemented per agreements with the individual State regulators. Upon establishing an agreement with the individual States, Tribal Nations, and other stakeholders, long-term surveillance and monitoring of the subsurface is assumed in perpetuity and planned for 100 years. Exhibit E-23 provides a summary of the currently assumed site end states for sites being managed by the Nevada Operations Office.

Exhibit E-23

Summary of Nevada Operations Office End States

Site Name	End State Description
Nevada Test Site	Decisions regarding future land use on the NTS are awaiting the completion of the Resource Management Plan, which is scheduled for October 1998. Surface soil plumes that straddle or extend outside the NTS boundaries will be characterized and remediated. Soil areas within the boundaries of the site will be characterized and monitored. Subsurface contaminants in and around the underground shot cavities will not be remediated since cost-effective remediation technologies have not yet been demonstrated. All of the site will remain under controlled access, however economic redevelopment is possible for the southwestern portion of the site. TRU and mixed TRU will be characterized and shipped to WIPP. On-site MLLW will be treated and disposed of on or off site. Environmental Restoration generated MLLW will be disposed of. LLW from approved generators on and off site will be disposed of in Area 3 and Area 5 of the Nevada Test Site. Filled disposal pits and trenches will be closed and capped according to approved closure designs and plans.
Tonopah Test Range	The site is currently part of the Nellis Air Force Range and the Department of Defense is responsible for the site future use. Soil hot-spots will be removed and cleaned to levels agreed upon with the State. Contamination in the industrial areas at the site will be closed in place and covered with engineered caps. The site is expected to remain under controlled access.
Amchitka Island, Central Nevada Test Area, Project Gas Buggy, Project Gnome-Coach, Project Rio Blanco, Project Rulison, Project Salmon, Project Shoal	DOE will not maintain an active presence at these sites. It is currently anticipated that following completion of all remedial activities, the surface areas will be released for alternate uses. However, it is also anticipated that the Department of Energy will maintain subsurface restrictions (institutional control) on all subsurface areas in proximity to the shot cavities.

E.5.2 Cost and Completion Dates

Nevada Operations Office has divided its environmental management work into 10 discrete projects comprising six environmental restoration projects and four waste management projects. A Project Baseline Summary (PBS) exists for each project and contains detailed programmatic information, including cost, schedule, scope, end state, and interim milestones. A summary of the Nevada Operations Office cost and schedule information is illustrated in Exhibit E-24. Although the Nevada Test Site EM mission is scheduled for completion in 2014, NTS will be open to receive low-level waste from other sites through 2070. For additional information on these projects, refer to individual PBSs. The overall planned site restoration completion dates are as follows:

Site	Date
Nevada Test Site	2014
Amchitka Island	2001
Central Nevada Test Site	2006
Gasbuggy	2005
Gnome-Coach	2004
Rio Blanco	2005
Rulison	1998
Salmon Site	1999
Shoal	2004
Tonopah Test Range	2007

The estimated EM life-cycle cost of Nevada Operations Office site cleanup is $2.2 billion (constant 1998 dollars) with environmental restoration ending in 2014, and waste management for low-level waste disposal activities ending in 2070. Long-term surveillance and monitoring will continue after restoration land disposal activities are complete.

Exhibit E-24 Nevada Operations Office
Cleanup Project Summary: Duration and Costs (All costs in thousands of 1998 dollars)

Site Closure Project Activities	1997 - 2006	2007 - 2070	Total
Nevada Off-sites			
Off-sites Remedial Action	65,324	24,946	90,270
Nevada Test Site			
Transuranic/Mixed Transuranic	32,857	0	32,857
Soils	180,523	22,412	202,935
Mixed Low-Level Waste	5,665	388	6,053
Industrial Sites	186,522	132,670	319,192
Underground Test Area (UGTA)	246,135	247,561	493,696
Low-Level Waste	84,535	589,612	674,147
Program Integration	69,260	106,615	175,875
Agreements in Principle/Grants	22,994	3,721	26,715
Program Management	29,610	187,998	217,608
Total	**923,426**	**1,315,923**	**2,239,349**

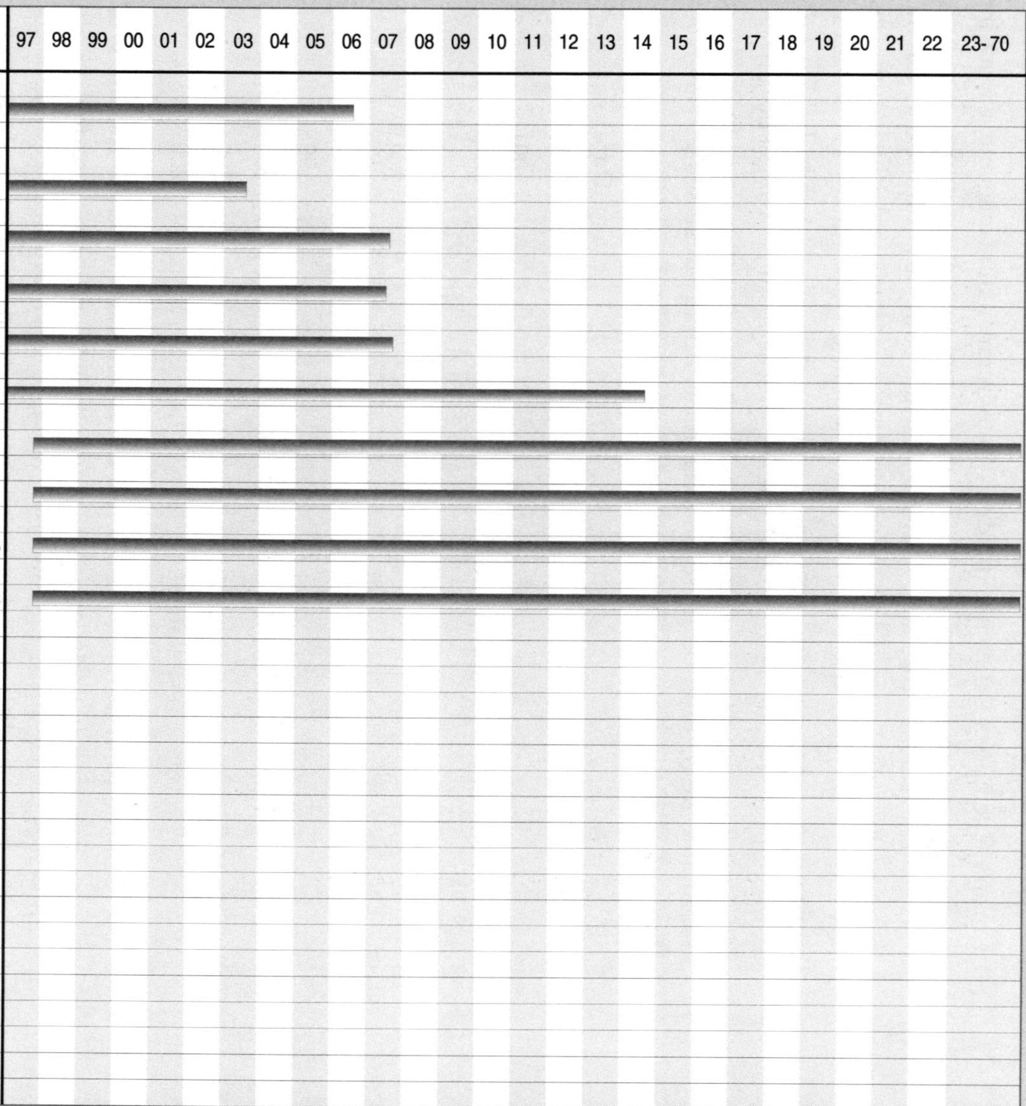

The projected cost profile for environmental management associated with the Nevada Operations Office is developed by combining the cost estimates in each of the Project Baseline Summaries. Exhibit E-25 displays the resultant baseline cost profile.

Exhibit E-25
Nevada Operations Office Environmental Management Baseline Cost Profile

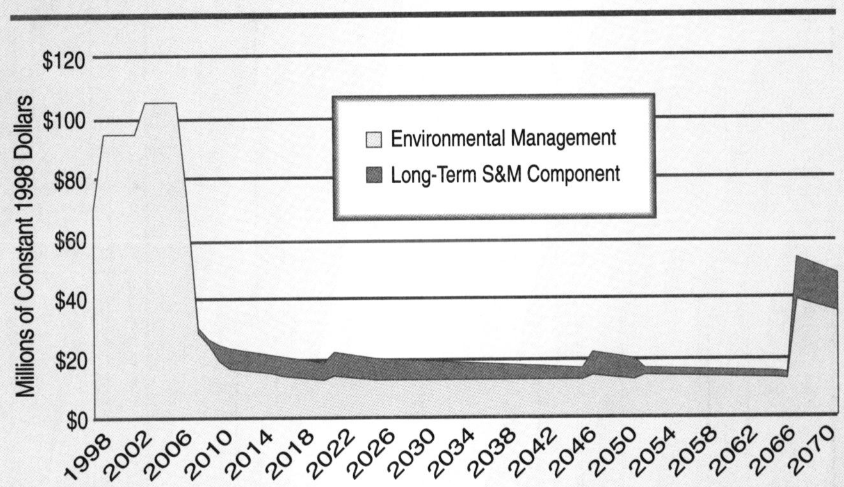

E.5.3 Work Scope Summary

The Environmental Management program at the Nevada Test Site consists of three divisions: Environmental Restoration, Waste Management, and Technology Development. Each division ensures that all federal laws and regulations are followed at DOE sites in the process of investigation, remediation, handling, transportation, disposal, and monitoring of the contaminated materials generated through weapons testing activities. For purposes of this document, only two of the divisions will be discussed, Environmental Restoration and Waste Management. The sections below describe the scope of work at the Nevada Operations Office. The volumes reported are approximate, and correspond to the major waste, material, and media flows, potential treatment processes, and off-site disposal destinations presented in Exhibit E-26, the Nevada Operations Office Conceptual Summary Disposition Map.

Environmental Restoration

The Environmental Restoration division determines remedial solutions to areas contaminated by nuclear weapons testing activities. The environmental restoration process involves identifying the nature of the contamination,

Exhibit E-26

Nevada Operations Office Conceptual Summary Disposition Map

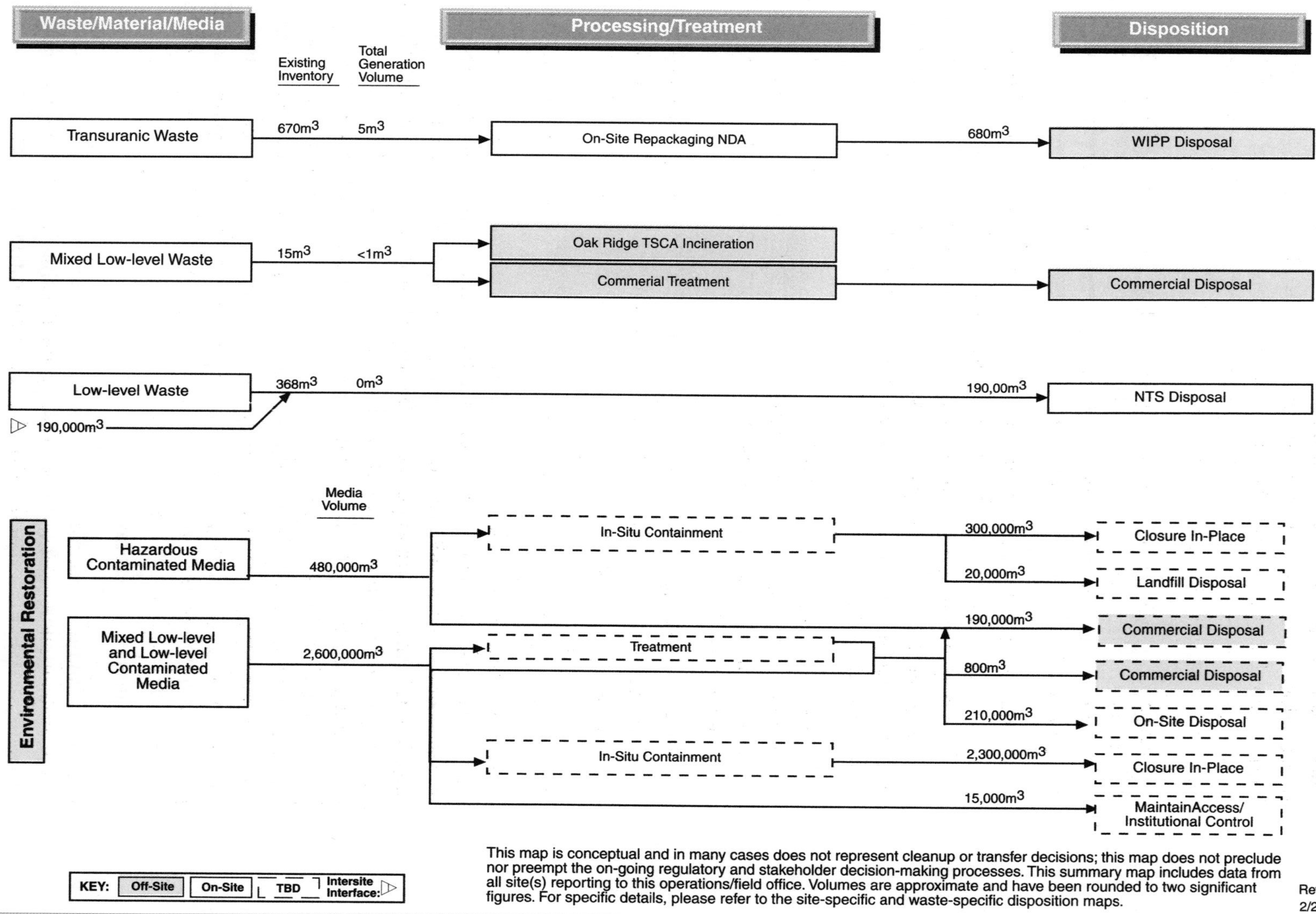

This map is conceptual and in many cases does not represent cleanup or transfer decisions; this map does not preclude nor preempt the on-going regulatory and stakeholder decision-making processes. This summary map includes data from all site(s) reporting to this operations/field office. Volumes are approximate and have been rounded to two significant figures. For specific details, please refer to the site-specific and waste-specific disposition maps.

Rev. 4.0
2/20/98

determining the risk to the public and the environment, acting to protect or restore the natural resources adversely affected by the releases of hazardous substances, and monitoring to ensure the safety of the site. Four main areas of remediation have been identified by the Nevada Operations Office: the Underground Test Areas (UGTA), the Industrial Sites, the Soil sites, and the Off-sites. The Nevada Operations Office also has projects for Program Integration and Agreements in Principle and Grants.

Underground Test Areas were contaminated by underground nuclear detonations above and within the water table. In order to ensure long-term health and safety, modeling and monitoring is conducted to predict movement of radionuclides in the groundwater.

Industrial Sites are areas contaminated with hazardous constituents from support activities for nuclear testing. These sites include discarded batteries, drums with diesel and gas, and old munition sites. Of the identified sites, many are easily remediated by simple removal actions, however, there are numerous sites that require more complex remedial action, and may result in the isolation of the contamination.

Soil Sites are those where atmospheric and near-surface nuclear tests were conducted resulting in the contamination of surface soil. The soil is characterized, removed, safely packaged, and disposed of at a NTS waste management site.

Off-sites are testing areas outside the NTS. The NTS is responsible for remediating off-site locations in Alaska, Colorado, Mississippi, Nevada, and New Mexico. Remediation at these sites ranges from the drainage and excavation of a pond to the removal of petroleum products, to the recapping of an underground test area, to the removal of radionuclide contaminated soil.

The volumes associated with NTS remediation include approximately 480,000 cubic meters of environmental media contaminated with hazardous substances, of which 300,000 cubic meters are expected to be closed in-place, 20,000 cubic meters are expected to be disposed of at an on-site landfill, and the remaining volume is expected to be disposed of at an off-site commercial hazardous facility. NTS remediation also includes approximately 2.6 million cubic meters of low-level and mixed low-level contaminated environmental media, of which 2.3 million cubic meters are expected to be closed in-place and 15,000 cubic meters are expected to be managed through access and institutional controls. An additional 800 cubic meters are expected to be disposed of at an off-site commercial facility and 210,000 cubic meters are expected to be disposed of on site.

Waste Management

Nevada Operations Office Waste Management activities are grouped into four projects: Transuranic and Mixed Transuranic, Mixed Low-level Waste,

Low-level Waste, and Program Management. Waste Management activities are designed to safely store and/or dispose of the waste generated by DOE activities throughout the complex. There are approximately 670 cubic meters of transuranic waste in inventory and five cubic meters expected to be generated over the life cycle of cleanup operations. After repackaging, approximately 680 cubic meters are expected to be shipped to WIPP in New Mexico for disposal. Mixed low-level waste generated on site will be treated and disposed of either on site or off site. Approximately 368 cubic meters of low-level waste and 15 cubic meters of mixed low-level waste are currently in inventory. Additionally, the Nevada Test Site expects to receive approximately 190,000 cubic meters of low-level waste from other DOE sites for disposal at the Nevada Test Site. Low-level waste received from approved generators currently identified in the Nevada Test Site Environmental Impact Statement Record of Decision will be disposed of at the Radioactive Waste Management Sites in Areas 3 and 5 on the Nevada Test Site.

Exhibit E-27 illustrates Nevada's environmental management costs by major work scope categories. Remedial action costs drive the overall cost for the environmental management program at the Nevada Operations Office.

Exhibit E-27
Nevada Operations Office
Life-Cycle Costs by Category

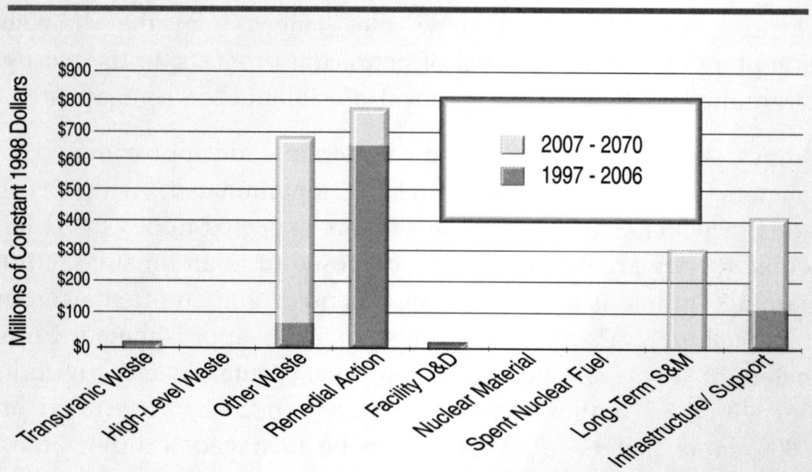

E.5.4 Critical Closure Path and Programmatic Risk

The critical closure path schedule presented as Exhibit E-28 sets forth the timetable for completing the closure activities at Nevada Operations Office. Completion of the EM mission at the Nevada Operations Office as scheduled will depend on the timely accomplishment of critical activities and milestones. In the

Exhibit E-28 Nevada Operations Office Critical Closure Path (Timelines and Risk Data)

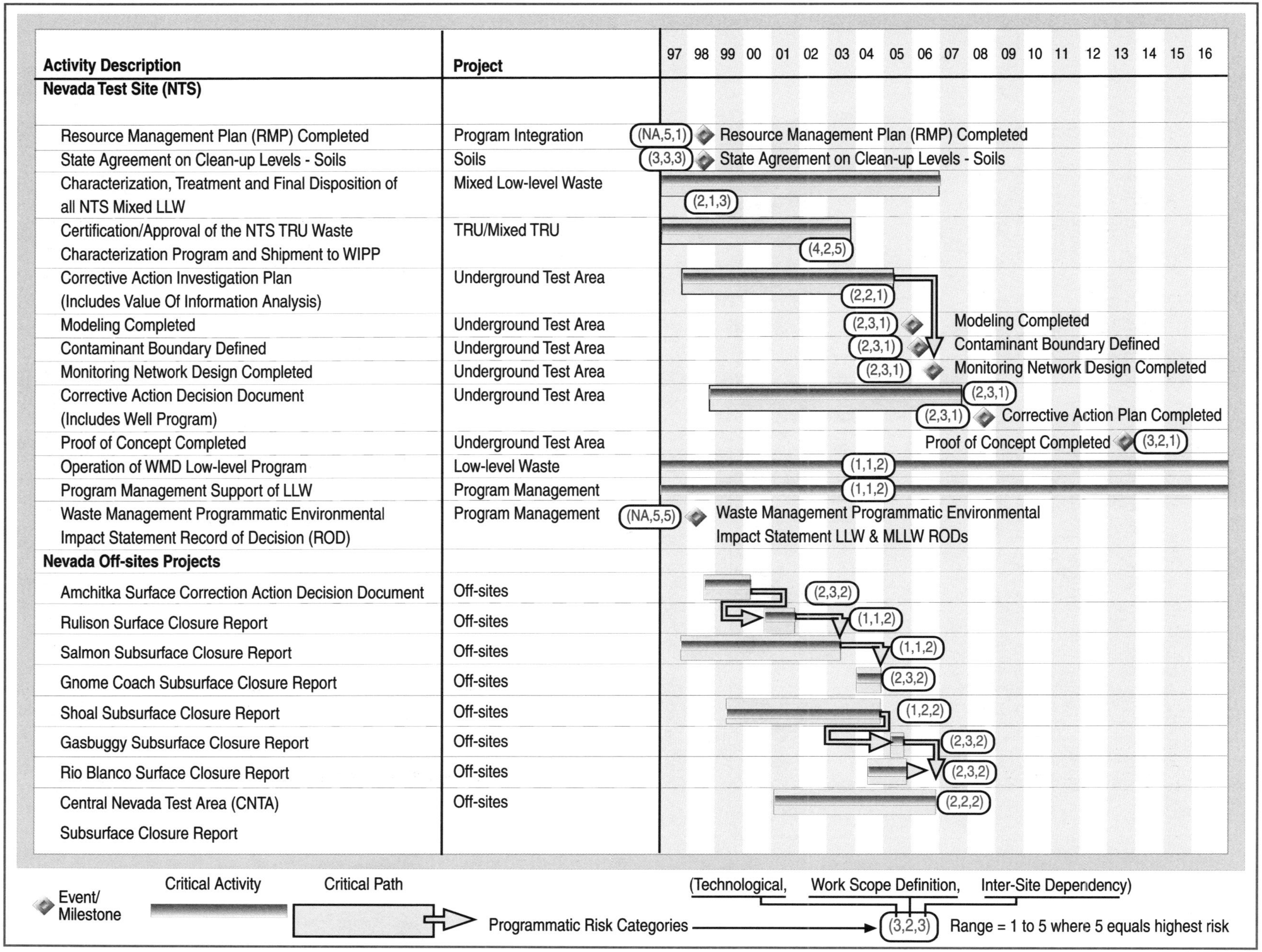

exhibit, the bars represent critical activities, while the diamonds represent events/milestones. Sites have assigned programmatic risk scores to each of the critical activities/milestones.

Appendix D provides a complete definition of programmatic risk. Exhibit E-29 presents a summary of activities/milestones on the critical closure path that have high programmatic risk (programmatic risk scores of 4 or 5 in any category). Nevada has three critical activities/events with high programmatic risk (i.e., risk scores of 4 or 5 in any category), including the certification and approval of the Nevada Test Site TRU waste characterization program and shipment of this waste to WIPP.

Exhibit E-29
Summary of High Programmatic Risk Activities/Milestones: Nevada Operations Office

Site	Project, Activity, Event	Start/End Dates	Programmatic Risk Categories		
			Technological	Work Scope Definition	Intersite Dependency
NTS	Waste Management Programmatic Environmental Impact Statement Records of Decision for LLW and MLLW	May 98	NA	5	5
	Resource Management Plan (RMP) completed	Oct 98	NA	5	1
	Certification/approval of the NTS TRU waste characterization program and shipment of waste to WIPP	Jan 96/ Sep 03	4	2	5

E.6 Oak Ridge Operations Office Summary

The mission of the Oak Ridge Operations Office is to oversee and manage various facilities and programs related to the Office of Nuclear Energy, Energy Research, Uranium Enrichment, Defense Programs, and Environmental Management in Tennessee, Ohio, Kentucky, and Missouri. The largest Oak Ridge Operations Office site, the Oak Ridge Reservation located in Oak Ridge, Tennessee, has approximately 1,100 acres of unlined radioactive and mixed waste burial grounds, inactive tanks, surplus facilities, and unlined ponds. As a result, soil, surface water, groundwater, and two major rivers in the area are contaminated. To address these issues and the issues at the Paducah Gaseous Diffusion Plant, the Portsmouth Gaseous Diffusion Plant, and the Weldon Spring Site, the Oak Ridge Operations Office has developed an aggressive strategy for the accelerated completion of its Office of Environmental Management mission.

Oak Ridge Operations Office

Three of the Oak Ridge Operations Office facilities are located on the Oak Ridge Reservation: the Oak Ridge National Laboratory, the Y-12 Plant; and the East Tennessee Technology Park. The Uranium Enrichment Gaseous Diffusion Plants in Paducah, Kentucky, and Portsmouth, Ohio, are also managed by the Oak Ridge Operations Office. Oak Ridge Operations Office is also responsible for the cleanup at the Weldon Spring Site in Missouri.

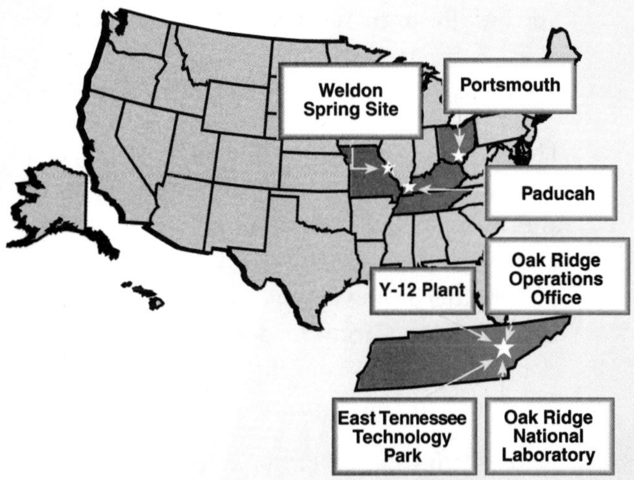

Oak Ridge National Laboratory is one of the country's largest multi-disciplinary and multi-program laboratories and research facilities. Weapons research facilities were established at the site of the Oak Ridge National Laboratory in 1943 as part of the World War II Manhattan Project. The laboratory's original mission was to produce and chemically separate the first gram quantities of plutonium as part of the national effort to produce the atomic bomb.

Y-12 Plant was built in 1943 as part of the Manhattan Project. The original mission of the Oak Ridge Y-12 Plant was a uranium enrichment and nuclear weapons production facility. Since World War II, the role of the Y-12 Plant has evolved into supporting highly sophisticated manufacturing; development engineering associated with the production, fabrication, and dismantlement of nuclear weapons components; and the national repository for enriched uranium.

The ***East Tennessee Technology Park*** (formerly K-25) was built as part of the Manhattan Project during World War II to supply enriched uranium for nuclear weapons production. From 1959 to 1969, the focus shifted to the production of commercial-grade, low-enriched uranium. Because of the declining demand for enriched uranium, the enrichment process was placed on standby in 1985 and shut down permanently in 1987. Currently, an effort is underway to industrialize ETTP by leasing facilities to private companies.

Construction of the ***Paducah*** and ***Portsmouth Gaseous Diffusion Plants*** began in the early 1950s to expand the federal government's gaseous diffusion program already in place at Oak Ridge, Tennessee. The facilities were built to increase the production of enriched uranium for defense and non-defense needs.

The ***Weldon Spring Site*** was part of a site used by the U.S. Army as an ordnance works in the 1940s. In the 1950s and 1960s, the Atomic Energy Commission used the site to process uranium ore in the Weldon Spring Chemical Plant. The plant was subsequently deactivated and no activities were carried out at the Weldon Spring Site until remediation began in 1985.

E.6.1 End State

The overall end state of the sites managed by the Oak Ridge Operations Office is assumed to be composed of some combination of controlled access, restricted and unrestricted industrial, and open space/recreational. An effort is currently underway to strengthen the end use recommendations through a process of stakeholder involvement. The Site-Specific Advisory Board has formed the End Use Working Group to develop end use assumptions that can be used to guide cleanup activities on the Oak Ridge Reservation. Actual end uses will be identified in the appropriate watershed or subproject Records of Decision.

At the Paducah and Portsmouth Gaseous Diffusion Plants and the Weldon Spring Site, discussions with the regulators and stakeholders will continue. The Paducah Gaseous Diffusion Plant continues to inform its Site-Specific Advisory Board concerning the prioritization and sequencing of work, and the Portsmouth Gaseous Diffusion Plant continues meeting with the U.S. Environmental Protection Agency and the Ohio Environmental Protection Agency.

Exhibit E-30 provides a summary of the anticipated site end states for Oak Ridge Operations Office.

Exhibit E-30
Summary of Oak Ridge Operations Office End States

Site Name	End State Description
Oak Ridge Reservation (ORR)	The Oak Ridge Reservation is comprised of the Oak Ridge National Lab (ORNL), the East Tennessee Technology Park (ETTP; formerly called K-25), and the Oak Ridge Y-12 Plant. Legacy waste stored at the ORR site will be disposed by 2006 for all transuranic waste, 2006 for all mixed low-level waste, and 2013 for all low-level waste. At ORNL, buried waste in both the Melton and Bethel Valleys will remain isolated in place with engineered and institutional controls implemented to prevent migration. Most contaminated media will be remediated in-situ, but hot spots and mercury contaminated soils will be excavated. Contaminated sediments in White Oak Creek, White Oak Lake, and White Oak Creek Embayment will be stabilized. Inactive buildings will be decontaminated and dismantled to grade except for the ORNL Graphite Reactor, which will be preserved as a national landmark. The Y-12 Plant will support restricted industrial, controlled access, and open space/recreational end uses. Burial grounds and other sources will be capped with contamination in place. Groundwater will be contained and use will be restricted. Some areas will be under controlled access for secure storage of nuclear materials and waste. The Environmental Management Waste Management Disposal Facility (EMWMDF) will be constructed on site for disposal of CERCLA waste. The ETTP end use is expected to be open space/ recreational, controlled access, and industrial with restrictions. The site is expected to be an industrial park occupied by private business. Contaminated areas within the reindustrialized area will be contained or consolidated. Selected facilities will be decontaminated and reused. Burial grounds will be capped and hydrologically isolated and/or excavated with waste disposed of at the EMWMDF or other appropriate disposal facility.
Paducah Gaseous Diffusion Plant	The gaseous diffusion process will remain operational, and the remaining property will be restricted industrial, open space/recreational, and controlled access. Several landfills or burial grounds will be closed with contamination remaining in place in the industrial area. Facilities will be cleaned for release or reuse, with deed restrictions or use limitations for areas with residual contamination. The off-site groundwater plumes will require long-term pump and treat operations to reduce migration and prevent discharges to surface water.

Exhibit E-30 (Continued)

Site Name	End State Description
Portsmouth Gaseous Diffusion Plant	The gaseous diffusion process will remain operational, and the remaining property will be restricted industrial, open space/recreational, and controlled access. Major sources of on-site contamination will be contained and/or remediated. Reindustrialization of existing DOE facilities is a possibility with deed restrictions or land use limitations on areas with contamination remaining in place. Several landfills or burial grounds will be closed with contamination remaining in place. Active groundwater treatment facilities will be shut down in 2050. Passive groundwater monitoring and treatment will continue until 2055.
Weldon Spring Site	155 acres of the Chemical Plant site will be released to the appropriate agency for unrestricted use, the 9-acre quarry will be released for recreational use, and the 62-acre on-site disposal cell will remain under controlled access.

E.6.2 Cost and Completion Dates

Oak Ridge Operations Office has divided its environmental management work into 28 discrete projects. A Project Baseline Summary (PBS) exists for each project and contains detailed programmatic information, including cost, schedule, scope, end state, and interim milestones. A summary of the Oak Ridge cost and schedule information is illustrated in Exhibit E-31. For additional information about these projects, see the Project Baseline Summaries.

The estimated EM life-cycle cost of Oak Ridge Operations Office site cleanups is $13.1 billion (constant 1998 dollars). The overall site planned completion dates are as follows:

Site	**Date**
Center for Energy and Environmental Research ...	1998
Oak Ridge Reservation	2013
Paducah Gaseous Diffusion Plant	2010
Portsmouth Gaseous Diffusion Plant	2005
Weldon Spring Site	2002

Exhibit E-31 Oak Ridge Operations Office
Cleanup Project Summary: Duration and Costs (All costs in thousands of 1998 dollars)

Site Closure Project Activities	1997 - 2006	2007 - 2070	Total	97–23-70
Oak Ridge Operations Office				
Directed Support	62,151	0	62,151	
Oak Ridge Reservation				
Sanitary/Industrial Waste Management	96,231	393,452	489,683	**Project in Steady State**
Hazardous Waste Management	58,394	280,395	338,789	**Project in Steady State**
East Tennessee Technology Park (ETTP) Facility Safety Upgrades	37,955	0	37,955	
Nuclear Materials and Facility Stabilization	54,769	0	54,769	
Transuranic Waste Privatization	110,195	0	110,195	
On-site Waste Management Facility	74,452	0	74,452	
Off-site Remedial Action	613,656	413,390	1,027,046	
Mixed Low-level Waste Management	660,747	2,197,553	2,858,300	
ETTP Landlord	115,333	0	115,333	
Transuranic Waste Management	237,446	21,303	258,749	
ETTP Process Equipment Decontamination and Decommissioning	447,411	66,522	513,933	
ETTP Decontamination and Decommissioning	232,322	70,264	302,586	
Y-12 Bear Creek Remedial Action	82,669	12,873	95,542	
Y-12 East Fork Poplar Creek Remedial Action	119,904	287,475	407,379	

Exhibit E-31 Oak Ridge Operations Office (Continued)
Cleanup Project Summary: Duration and Costs (All costs in thousands of 1998 dollars)

Site Closure Project Activities	1997 - 2006	2007-2070	Total
ORNL Bethel Valley Decontamination and Decommissioning Project	69,907	111,563	181,470
ETTP Remedial Action	145,486	413,057	558,543
Oak Ridge National Laboratory (ORNL) Bethel Valley Remedial Action	145,449	484,313	629,761
ORNL Melton Valley Watershed Decontamination and Decommissioning	103,992	49,156	153,148
ORNL Melton Valley Watershed Remedial Action	222,437	228,448	450,885
Low-level Waste Management	484,500	1,833,415	2,317,916
Paducah Gaseous Diffusion Plant			
Paducah Waste Management	237,405	11,462	248,868
Paducah Remedial Action	220,301	432,461	652,762
Portsmouth			
Portsmouth Remedial Action	155,794	402,554	558,348
Portsmouth Waste Management	273,022	4,064	277,086
Weldon Spring			
Weldon Spring Waste Treatment	70,737	0	70,737
Weldon Spring Disposal Facility	266,072	0	266,072
Weldon Spring Long-term Surveillance and Maintenance	5,149	22,836	27,985
Total	**5,403,886**	**7,736,555**	**13,140,442**

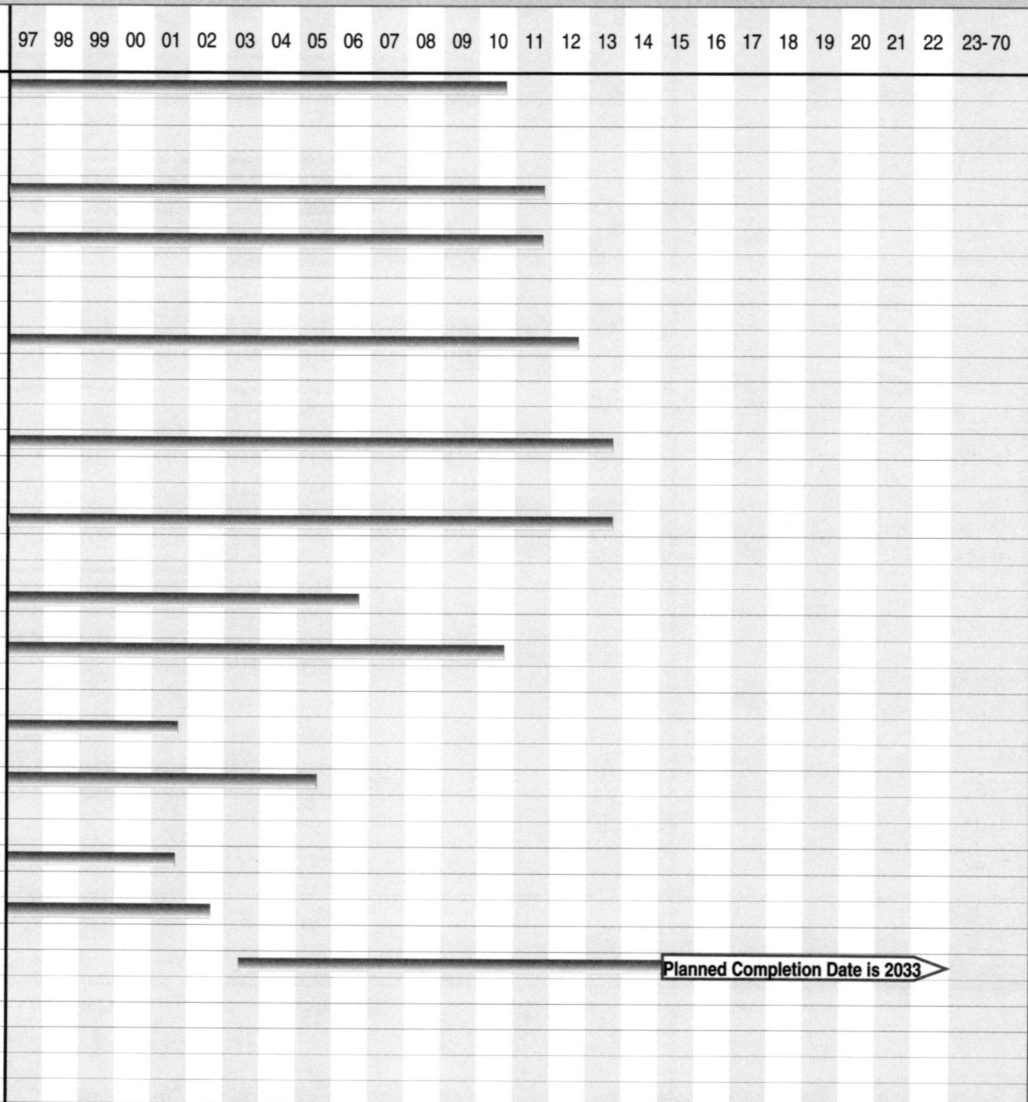

The projected cost profile for environmental management associated with the Oak Ridge Operations Office is developed by combining the cost estimates in each of the PBS. Exhibit E-32 displays the resultant baseline cost profile.

Exhibit E-32
Oak Ridge Operations Office
Environmental Management Baseline Cost Profile

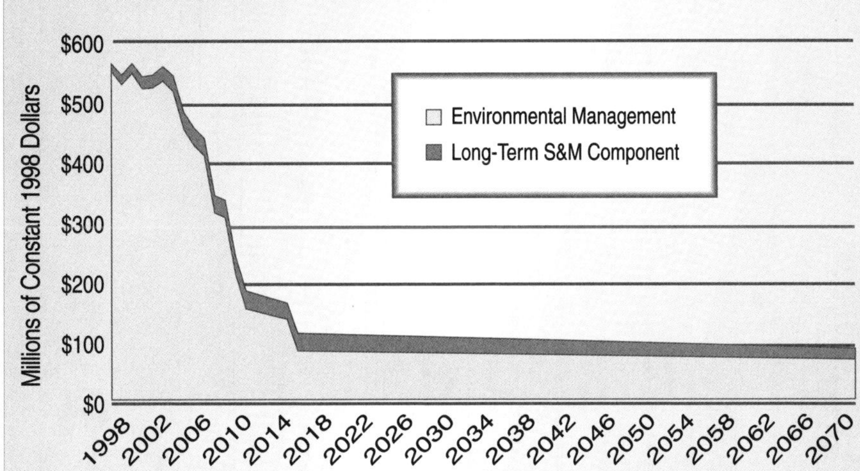

E.6.3 Work Scope Summary

The scope of work at the Oak Ridge Operations Office encompasses the Oak Ridge Reservation, Portsmouth and Paducah Gaseous Diffusion Plants, and the Weldon Spring Site. Cleanup activities at these sites include the management of depleted uranium and spent nuclear fuel; treatment of off-site mixed low-level waste at the Toxic Substances Control Act (TSCA) incinerator; and disposal of contaminated soils and debris at the Weldon Spring disposal cell. Cleanup activities include the deactivation of 33 facilities, the decommissioning of 401 facilities, and the cleanup of 642 release sites. The sections below describe the major waste, material, and contaminated media volumes to be addressed by the Oak Ridge Operations Office. The volumes reported are approximate, and correspond to the major waste, material, and media flows, potential treatment processes, and off-site disposal destinations presented in Exhibit E-33, the Oak Ridge Operations Office Conceptual Summary Disposition Map.

Exhibit E-33 Oak Ridge Operations Office Conceptual Summary Disposition Map

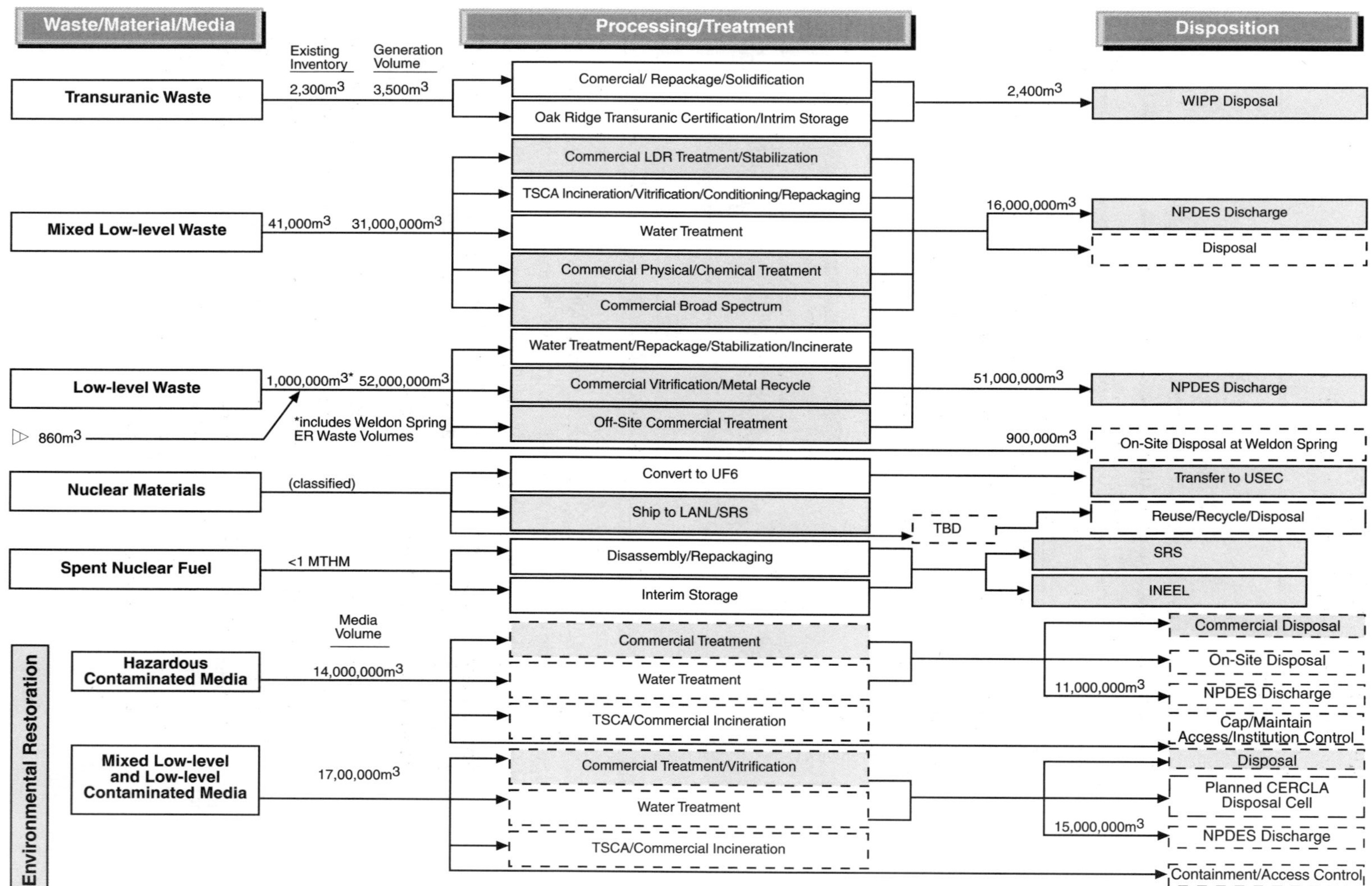

This map is conceptual and in many cases does not represent cleanup or transfer decisions; this map does not preclude nor preempt the on-going regulatory and stakeholder decision-making processes. This summary map includes data from all site(s) reporting to this operations/field office. Volumes are approximate and have been rounded to two significant figures. For specific details, please refer to the site-specific and waste-specific disposition maps.

Rev. 4.0
2/20/98

Transuranic Waste

- Approximately 2,300 cubic meters of transuranic waste are currently in inventory and 3,500 cubic meters of transuranic waste are expected to be generated over the life cycle of operations. After treatment and repackaging, 2,400 cubic meters are expected to be disposed of at the Waste Isolation Pilot Plant (WIPP).

Other Waste

- Approximately 41,000 cubic meters of mixed low-level waste are currently in inventory and nearly 31 million cubic meters of solid and liquid low-level waste are expected to be generated over the life cycle of operations. After undergoing a range of treatment activities, 16 million cubic meters of treated effluent will be discharged under an NPDES permit, and an additional amount of solid waste is expected to be disposed of at an undetermined facility.

- In addition to one million cubic meters of low-level waste that are currently in inventory, 860 cubic meters of low-level waste are expected to be transferred from other sites and 52 million cubic meters of low-level waste waters and liquids are expected to be generated over the life cycle of operations. After treatment, 51 million cubic meters of treated effluent will be discharged under an NPDES permit, and an additional 900,000 cubic meters are expected to be directly disposed of on site at Weldon Spring.

Remedial Action and Facility D&D

- Remedial Action: Over 14 million cubic meters of environmental media including solids, sludge, and debris and groundwater contaminated with hazardous substances are planned to be managed. Media will undergo a range of treatment activities including off-site commercial treatment. After treatment, 11 million cubic meters of effluent will be discharged under an NPDES permit and undetermined volumes are expected to be disposed of on-site and at an off-site commercial facility. An additional undetermined volume will be capped in place and maintained under access and institutional control.

- Facility D&D: Over 17 million cubic meters of contaminated environmental media including soils, sludges, debris, and groundwater contaminated with radionuclides and hazardous substances are planned to be managed. Media will undergo a range of treatment including off-site commercial incineration. After treatment, 15 million cubic meters of treated effluent will be discharged under an NPDES permit, and undetermined volumes are expected to be disposed of in the EMWMDF and an undetermined facility. An additional undetermined volume is expected to be contained in place and maintained under access control.

Nuclear Materials

- Quantities of the following materials for this program are sensitive and cannot be disclosed in this document. Classified volumes of plutonium and uranium metals, oxides, and solutions will be managed; some will be converted to UF6 and transferred to the United States Enrichment Corporation; remaining volumes will be transferred to other DOE sites for reuse, recycling, or disposal.

Spent Nuclear Fuel

- Less than one metric ton of spent nuclear fuel will be managed. After disassembly and repackaging, spent nuclear fuel will be transferred to the Savannah River Site and the Idaho National Engineering and Environmental Laboratory.

Exhibit E-34 displays site closure costs at the Oak Ridge Operations Office by major work scope category.

Exhibit E-34
Oak Ridge Operations Office
Life-Cycle Costs by Category

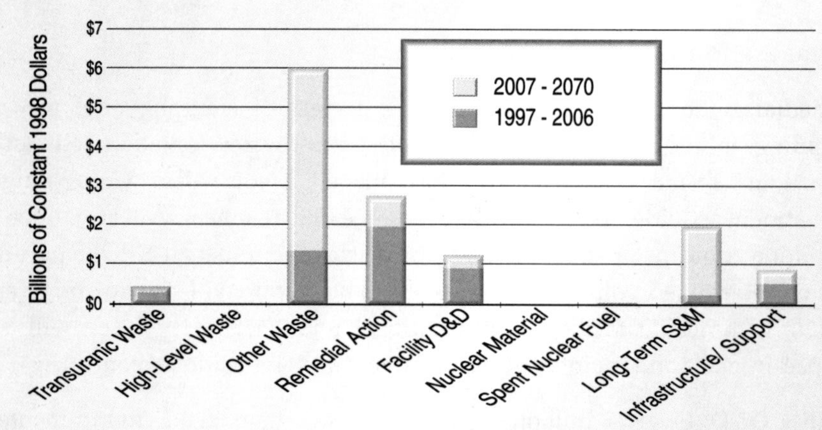

E.6.4 Critical Closure Path and Programmatic Risk

The critical closure path schedule presented as Exhibit E-35 sets forth the timetable for completing the closure activities at Oak Ridge Operations Office. The highlighted activities show the critical closure path, which represents the series of events that drive the overall completion date for the site. In Exhibit E-35, the bars represent critical activities, and the diamonds represent milestones and critical events.

Exhibit E-35 Oak Ridge Operations Office Critical Closure Path (Timelines and Risk Data)

Activity Description	Project
Oak Ridge Reservation (ORR)	
Private Sector Contract Award for Treatment and Disposal of Mixed Waste Stored on the ORR	
Treat Inventory of Mixed Low-Level Waste	Mixed Low Level Waste Management
Disposition of Legacy Low-Level Waste	Low Level Waste Management
RH-TRU Contract Award for Collection, Stabilization & Containerization of ORNL RH-TRU Waste to meet WIPP WAC	Transuranic Waste Privatization
Transuranic Waste Steady State	Transuranic Waste Management
Construction of EMWMDF	On-site Waste Management Facility
UEFPC ROD for Multiple Contaminant Sources and Commingled Surface and Groundwater	Y-12 East Fork Poplar Creek Remedial Action
UEFPC Soil Remediation	Y-12 East Fork Poplar Creek Remedial Action
Bear Creek Valley ROD for Multiple Contaminant Sources, Groundwater & Surface Water West of Y-12 Plant	Y-12 Bear Creek Remedial Action
Bear Creek Valley Boneyard/Burnyard	Y-12 Bear Creek Remedial Action
Bear Creek Valley Groundwater Remediation	Y-12 Bear Creek Remedial Action
SWSA 5 North and South Remediation	ORNL Melton Valley Watershed Remedial Action
White Oak Creek Remediation	ORNL Melton Valley Watershed Remedial Action
Record of Decision for Contaminated Areas within the Melton Valley Watershed	ORNL Melton Valley Watershed Remedial Action

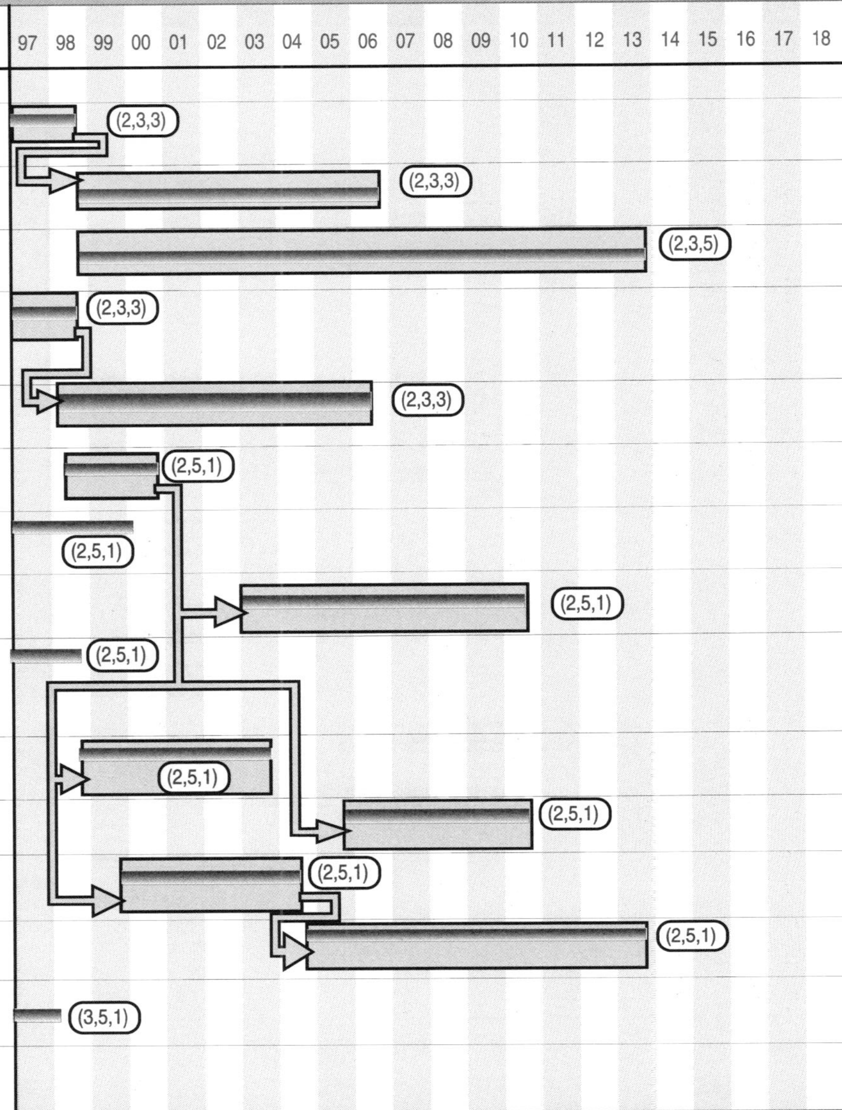

Event/Milestone | Critical Activity | Critical Path

Programmatic Risk Categories → (3,2,3) (Technological, Work Scope Definition, Inter-Site Dependency) Range = 1 to 5 where 5 equals highest risk

Exhibit E-35 **Oak Ridge Operations Office Critical Closure Path (Continued)** (Timelines and Risk Data)

Activity Description	Project	97 98 99 00 01 02 03 04 05 06 07 08 09 10 11 12 13 14 15 16 17 18
Record of Decision for Contaminated Areas within the Bethel Valley Watershed	ORNL Bethel Valley Remedial Waste Management	(2,5,1)
East Tennessee Technology Park (ETTP) ROD for Contaminated Areas, Groundwater and Surface Water	ETTP Remedial Action	(2,5,1)
K-29, K-31, K-33 Process Equipment Deactivation & Decommissioning	ETTP Process Equipment D&D	(2,2,1)
K-25, K-27 Process Equipment Deactivation & Decommissioning	ETTP Process Equipment D&D	(2,2,1)
K-25/27/29 Building Demolition	ETTP D&D	(2,5,1)
Paducah Gaseous Diffusion Plant (PGDP)		
Complete Removal of UF4 Scrap Pile	Paducah Remedial Action	(1,3,3)
Complete Sources of Off-site Contamination (WAG 3) -- Investigate & Remediate 3 Burial Grounds	Paducah Waste Management	(1,5,1)
Shipment of LDR Mixed Waste Solids to Toxic Substances Control Act (TSCA) Incinerator	Paducah Waste Management	(1,3,3)
Cleanup Groundwater and Surface Water Units	Paducah Remedial Action	(2,5,1)
Shipment of Cyanide Waste for Treatment	Paducah Waste Management	(2,1,3)
Treatment of Mixed Wastes at C-400-D	Paducah Waste Management	(1,1,1)
Complete Site Evaluations of Low Risk WAGS (WAG 30), 8 Release Sites	Paducah Waste Management	(1,5,1)
Complete Toxicity Characteristic Leaching Procedure (TCLP) per Federal Facility Compliance Agreement (FFCA)	Paducah Waste Management	(1,1,1)
Shipment of Photographic Waste for Treatment	Paducah Waste Management	(2,1,3)
Shipment of Commercial Stabilization Waste for Treatment	Paducah Waste Management	(3,1,3)
Shipment of LDR Mixed Waste Liquid to TSCA Incinerator	Paducah Waste Management	(1,3,3)

Event/ Milestone — Critical Activity — Critical Path

Programmatic Risk Categories → (3,2,3) (Technological, Work Scope Definition, Inter-Site Dependency) Range = 1 to 5 where 5 equals highest risk

Exhibit E-35 Oak Ridge Operations Office Critical Closure Path (Continued) (Timelines and Risk Data)

Activity Description	Project	Risk (T,W,I)
Portsmouth Gaseous Diffusion Plant		
Quadrant I, II, III, IV Decision Document	Portsmouth Remedial Action	(1,3,1)
Complete Quadrant I, II, III, and IV CMI	Portsmouth Remedial Action	(1,1,1)
Complete Remedial Action Cleanup	Portsmouth Remedial Action	(2,3,1)
Complete X-701B CMI	Portsmouth Remedial Action	(2,3,1)
Reactive Media Selection - X-701B GW	Portsmouth Remedial Action	(2,3,1)
Reactive Media Selection - X-749/X-120	Portsmouth Remedial Action	(2,3,1)
X-701B Decision Document	Portsmouth Remedial Action	(1,3,1)
X-749/X-120 Decision Document	Portsmouth Remedial Action	(1,3,1)
Legacy Waste Treatment and Disposal	Portsmouth Remedial Action	(2,2,3)
X-701B Area	Portsmouth Remedial Action	(2,3,2)
Peter Kiewit Landfill Cap	Portsmouth Remedial Action	(1,1,1)
Complete Low-Level Waste Characterizations	Portsmouth Waste Management	(1,2,1)
Complete TSCA Incineration	Portsmouth Waste Management	(2,1,3)
Complete Waste Water Treatment	Portsmouth Waste Management	(1,1,1)
Complete Storage Project	Portsmouth Waste Management	(2,3,3)
Resource Conservation and Recovery Act (RCRA) Characterizations	Portsmouth Waste Management	(1,3,1)
Complete TSCA Disposal	Portsmouth Waste Management	(2,3,3)
Complete Low-Level Waste Treatment & Disposal	Portsmouth Waste Management	(2,2,3)
Complete RCRA Treatment	Portsmouth Waste Management	(2,2,3)
Weldon Spring Site		
Records Of Decision (RODs) for Site Groundwater and Quarry Residuals	Weldon Spring Disposal Facility	(2,3,2)
Weldon Spring Disposal Cell Operation	Weldon Spring Disposal Facility	(2,2,2)
Disposal Cell Construction	Weldon Spring Disposal Facility	(2,2,2)
Raffinate Pit Remediation	Weldon Spring Waste Treatment	(1,1,1)

Timeline: 97 98 99 00 01 02 03 04 05 06 07 08 09 10 11 12 13 14 15 16 17 18

Event/Milestone; Critical Activity; Critical Path

Programmatic Risk Categories (Technological, Work Scope Definition, Inter-Site Dependency) → (3,2,3) Range = 1 to 5 where 5 equals highest risk

Completion of the EM mission at Oak Ridge Operations Office, as scheduled, will depend on the timely accomplishment of critical activities and milestones. Sites have assigned programmatic risk scores to each of the critical activities/milestones. Appendix D provides a complete definition of programmatic risk. Exhibit E-36 presents a summary of activities/milestones on the critical closure path that have high programmatic risk (programmatic risk scores of 4 or 5 in any category). For cleanup activities, the major uncertainties are in the definition of work scope. Cleanup actions are assumed and may change after the approval of decision documents. For certain waste management activities, disposal location is uncertain which results in a high programmatic risk score. The high programmatic risk will decrease after the disposal agreements are reached. The Oak Ridge Operations Office version of *Paths to Closure* provides more details on the management approach for these high programmatic risk issues.

Exhibit E-36

Summary of High Programmatic Risk Activities/Milestones: Oak Ridge Operations Office

Site	Project, Activity, Event	Start/End Dates	Programmatic Risk Categories		
			Technological	Work Scope Definition	Intersite Dependency
ORR	Record of Decision (ROD) for contaminated areas in the ORNL Melton Valley within the Melton Valley Watershed	Oct 96/ Jun 98	2	5	1
	Bear Creek Valley ROD for multiple contaminant sources, groundwater & surface water west of the Y-12 Plant	Oct 96/ Oct 98	2	5	1
	Bethel Valley ROD for contaminated areas in the Bethel Valley Watershed	Oct 96/ Apr 99	2	5	1
	ETTP ROD for contaminated areas, groundwater and surface water	Oct 96/ May 00	2	5	1
	UEFPC ROD for multiple contaminant sources and commingled surface and groundwater	Oct 96/ Feb 00	2	5	1
	Construction of Environmental Management Waste Management Disposal Facility	Oct 98/ Sep 00	2	5	1
	Bear Creek Valley Boneyard/ Burnyard	Oct 98/ Sep 03	2	5	1

Exhibit E-36 (Continued)

Summary of High Programmatic Risk Activities/Milestones: Oak Ridge Operations Office

Site	Project, Activity, Event	Start/End Dates	Programmatic Risk Categories		
			Technological	Work Scope Definition	Intersite Dependency
	SWSA 5 North and South Remediation	Oct 99/ Sep 04	2	5	1
	K-25/27/29 Building Demolition	Oct 04/ Oct 09	2	5	1
	Bear Creek Valley Groundwater Remediation	Oct 05/ Sep 10	2	5	1
	UEFPC Soil Remediation	Oct 02/ Sep 10	2	5	1
	Disposition of legacy LLW	Sep 98/ Sep 13	2	3	5
	White Oak Creek Remediation	Oct 04/ Sep 13	2	5	1
Paducah	Complete Sources of Off-Site Contamination	Jan 99/ Sep 03	1	5	1
	Complete site evaluations of low risk WAGS (WAG 30), 8 release sites	Oct 96/ Sep 04	1	5	1
	Cleanup groundwater and surface water units	Oct 03/ Sep 10	2	5	1

E.7 Oakland Operations Office Summary

The Oakland Operations Office oversees a wide range of programs and nine sites throughout California and one in New York State. Oakland's mission is to manage risks at these multiple research facilities which are contaminated with various hazardous and radioactive materials. The Office of Environmental Management (EM) activities at each of these sites vary. However, Oakland plans to have all EM missions completed at all sites (excluding the Separations Process Research Unit) by 2006. After the EM mission is complete, most sites have ongoing research missions that will be managed by the owner, however, the decision regarding the management of newly-generated waste is still pending.

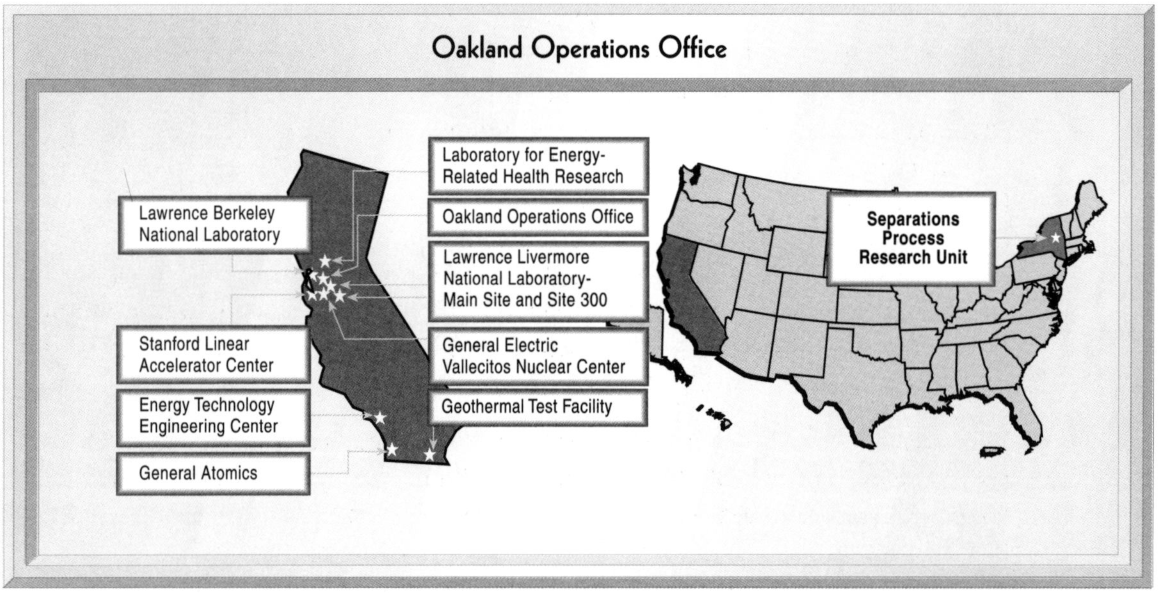

Energy Technology Engineering Center (ETEC) is located in the Simi Hills of Ventura County, approximately 30 miles northwest of downtown Los Angeles. The Energy Technology Engineering Center consists of government-owned buildings that occupy 90 acres owned by Boeing North American, Rocketdyne Division on the Santa Susana Field Laboratory. ETEC was established in the mid-1960s as a Department of Energy (DOE) laboratory to support nuclear research and energy development projects. All nuclear-related research ended by 1989. Office of Nuclear Energy activities at ETEC were terminated at the end of 1995. At ETEC the EM cleanup mission is focused primarily on remediating contaminated groundwater and soils in addition to the decontamination and decommissioning (D&D) of several buildings.

General Atomics (GA) occupies two contiguous sites that are located approximately 13 miles north of downtown San Diego. The overall mission of the EM program at General Atomics is the decontamination and demolition of the Hot Cell Facility. The Hot Cell Facility, which General Atomics owns and operates, has been used for numerous post-irradiation examinations of Department fuels, structural materials, reactor dosimetry materials, and instrumentation.

General Electric Vallecitos Nuclear Center (GE) is a privately-owned commercial site where past DOE operations have been performed. Past DOE fuel examination activities were responsible for contaminating the General Electric Vallecitos Nuclear Center high-level Hot Cell #4 and the Emissions Spectrograph (Glovebox). EM activities at the General Electric Vallecitos Nuclear Center are limited to the cleanup of these two areas.

The cleanup mission at the ***Geothermal Test Facility (GTF)*** was completed in the first quarter of FY 1997.

Laboratory for Energy-Related Health Research (LEHR) is an inactive research facility where, for a period of 30 years, DOE and its predecessors funded radiation-related studies using animals. The research program, concluded in 1988, was conducted by the University of California at Davis (UCD). In 1990, DOE initiated site restoration activities with emphasis on facility decontamination and the removal of high risk radioactive sources. In 1994, the LEHR site, along with the UCD landfills and burial trenches, were added to the U.S. Environmental Protection Agency's National Priority List. Under the terms of an agreement between DOE and the University, DOE is responsible for the remediation of contaminated areas including domestic and septic tanks, burial trenches, dry wells, underground waste treatment facilities, leach fields, and about four acres of outside dog pen facilities.

Lawrence Berkeley National Laboratory (LBNL) occupies 134 acres adjacent to the Berkeley Campus of the University of California. In the early 1930s, the University of California leased land to DOE for construction of the Lawrence Berkeley National Laboratory where DOE conducted numerous research activities. Buildings were constructed for a wide variety of energy-related research activities, including nuclear and high-energy physics, accelerator research and development, materials research, geology, molecular biology, and biomedical research. EM activities at LBNL involve remediation of soil and groundwater contamination produced by those activities.

Lawrence Livermore National Laboratory (LLNL) is composed of two sites, the Main Site and Site 300, both located approximately 50 miles east of San Francisco. DOE and the University of California jointly operate both sites. The Livermore Main Site was converted from agricultural use by the U.S. Navy in 1942 as a flight training base and for aircraft assembly, repair, and overhaul. In 1952 the site was transferred to the Atomic Energy Commission (AEC). Under AEC, the site became a weapons design and basic physics laboratory and continues with this mission under DOE today. Initial releases of hazardous materials occurred at the Livermore Site in the 1940s when the site was a Naval Air Station. There is also evidence that localized spills, leaking tanks, and impoundments and landfills contributed volatile organic compounds (VOCs), fuel hydrocarbons (FHCs), metals, and tritium to groundwater and unsaturated sediments after the Navy era. The LLNL Main Site was added to the EPA's National Priority List (NPL) in 1987. The purpose of this project is to characterize existing contamination and to effectively remediate soil and groundwater.

Site 300 was placed on the NPL in 1990 principally because of high concentrations of trichloroethylene (TCE) in groundwater and two off-site TCE groundwater plumes. A Comprehensive Environmental Response, Compensation, and Liability Act (CERCLA) Federal Facility Agreement was negotiated between DOE/LLNL, EPA, the State Department of Toxic Substances Control, and the California Central Valley Regional Water Quality Control Board in 1992 for Site 300 and in 1998 for the Main Site.

Separations Process Research Unit (SPRU), located at the Knolls Site of the Knolls Atomics Power Laboratory (KAPL) near Schenectady, New York, is an inactive complex requiring decontamination and decommissioning. SPRU was a pilot plant used for developing the redox and purex processes for extracting both uranium and plutonium from irradiated fuel. As a result of this work conducted by the Materials Production Division of the AEC in the early 1950s, associated buildings and the surrounding ground became contaminated. The complex, in standby status since 1953, has been accepted into the EM program for decontamination and decommissioning of contaminated facilities and remediation of contaminated soils. Until such decommissioning activities begin, a surveillance and monitoring program is in place to ensure that the facility remains in a stable condition and that it does not present an unacceptable risk to the public, the environment, or the on-site work force.

Stanford Linear Accelerator Center (SLAC) is a high-energy research facility, established in 1962, which is owned and operated by Stanford University under contract to DOE. The Center's four major experimental facilities are the Linear Accelerator, the Positron Electron Project Storage Ring, the Stanford Positron Electron Asymmetric Ring, and the Stanford Linear Accelerator Center Linear Collider. The primary objective of SLAC's EM program is to clean up contaminated soils and groundwater and to return the land to the site landlord, the Office of Energy Research, by the end of FY 2000 for beneficial use.

E.7.1 End State

Exhibit E-37 provides a summary of the anticipated end states for the Oakland Operations Office sites.

Exhibit E-37

Summary of Oakland Operations Office End States

Site Name	End State Description
Energy Technology Engineering Center	Environmental Management is responsible for remediation. Remediation will be complete by FY 2006, and the site will be turned over to Boeing North American. All wastes are being shipped off site. End state use will probably be industrial.
General Atomics	The site is expected to be fully remediated by FY 2000. The Hot Cell Facility will be decontaminated and decommissioned, and the site will be released as NRC no-rad restriction. Soil cleanup limits are based on an industrial land use. All wastes are being shipped off site, some to INEEL. DOE maintains liability at the site until all of the waste is off of the site.
General Electric Vallecitos Nuclear Center	Remediation of this site is expected to be complete by 2005, at which time DOE will have no further obligations to General Electric. The hot cell will be turned over to GE, who plans on using it commercially, though a portion of the site will be zoned industrial.
Geothermal Test Facility	The site was completed in the first quarter of FY 1997, and was turned over to the Bureau of Land Management in 1997 for unrestricted use. The brine pond waste material was removed and disposed of off site. No long-term monitoring, surveillance, or maintenance is required. A NEPA categorical exclusion was issued in accordance with 10 CFR 1021, Appendix B 6.1.
Laboratory for Energy-Related Health Research	Site cleanup will be complete by 2002. Closure of the RCRA storage facility is expected to end by FY 2001. UC-Davis is responsible for a radioactive waste burial trench and three landfills that are on site. Post-closure monitoring will primarily be the responsibility of UC-Davis. The four buildings that DOE is responsible for will be released for unrestricted use. All waste will be shipped off site.
Lawrence Berkeley National Laboratory	LBNL has an ongoing mission with continued generation of hazardous, mixed, and radioactive wastes. A groundwater treatment system is expected to be in place by 2003. Clean closure of the Hazardous Waste Handling Facility (HWHF) will be completed in FY 1998, and a new HWHF was constructed in FY 1997. No soil remediation of the HWHF is expected. Currently, no definitive cleanup level has been established for tritium in groundwater.

Exhibit E-37 (Continued)

Site Name	End State Description
Lawrence Livermore National Laboratory - Main Site	LLNL expects to continue to occupy and conduct research at the Main Site indefinitely. Future land use is expected to be industrial. VOCs have contaminated groundwater sources on and off site. Remediation of the soil and groundwater is in progress. By 2006, all of the soil and groundwater treatment facilities will be operating. No solid waste disposal will occur on site. DOE will continue to own and manage the site.
Lawrence Livermore National Laboratory - Site 300	LLNL expects to continue to occupy and conduct research at Site 300 indefinitely. Groundwater treatment systems will be in place and operational by FY 2006. Access will continue to be controlled. The land will continue to be a mix of industrial and wildlife areas. No solid waste disposal will occur on site.
Separations Process Research Unit	All radiological and hazardous contamination (LLW, MLLW, TRU, MTRU, HLW) will be disposed of off site. The majority of cleanup activities will occur between 2006 and 2014. The area is expected to be released for unrestricted use by the owner, Knolls Atomic Power Laboratory.
Stanford Linear Accelerator Center	This site has an ongoing mission as an active research facility. Cleanup of the contaminated areas will be completed by Environmental Restoration and the site returned to the Office of Energy Research by 2000. A network of wells has been installed to monitor groundwater contamination. Long-term monitoring responsibilities will be transferred to the site landlord, the Office of Energy Research. Contaminants will remain in the soil at depths of 10 to 20 feet near the Former Solvent Underground Storage Tank Area.

E.7.2 Cost and Completion Dates

Oakland Operations Office has divided its EM work into 21 discrete projects. A Project Baseline Summary (PBS) exists for each project and contains detailed programmatic information, including cost, schedule, scope, end state, and interim milestones. A summary of the Oakland cost and schedule information is illustrated in Exhibit E-38. For additional information about these projects, refer to the individual PBSs.

The estimated life-cycle cost of Oakland Operations Office environmental management work scope is $1.0 billion (constant 1998 dollars). This estimate does not include approximately $1.1 billion (constant 1998 dollars) in costs associated with the generation of new wastes that are expected to be the responsibility of the generator.

Exhibit E-38 Oakland Operations Office
Cleanup Project Summary: Duration and Costs (All costs in thousands of 1998 dollars)

Site Closure Project Activities	1997 - 2006	2007 - 2070	Total
Oakland Operations Office			
State Grants	3,402	1,933	5,335
Environmental Technology Engineering Center (ETEC)			
ETEC Landlord	15,989	47,312	63,300
ETEC Waste Management	38,385	0	38,385
ETEC Remediation	127,607	0	127,607
General Atomics			
Hot Cell Facility Deactivation and Decommissioning at General Atomics	11,477	0	11,477
General Electric			
General Electric Deactivation and Decommissioning (Environmental Restoration)	20,650	0	20,650
Geothermal Test Facility			
Soil Remediation at the Geothermal Test Facility	1,335	0	1,335
Laboratory for Energy Related Health Research (LEHR)			
LEHR Waste Management	4,272	0	4,272
LEHR Environmental Restoration	17,727	0	17,727
Lawrence Berkeley National Laboratory (LBNL)			
LBNL Hazardous Waste Handling Facility Closure (Environmental Restoration)	1,136	0	1,136
LBNL Soils and Groundwater (Environmental Restoration)	25,3545	27,684	53,038

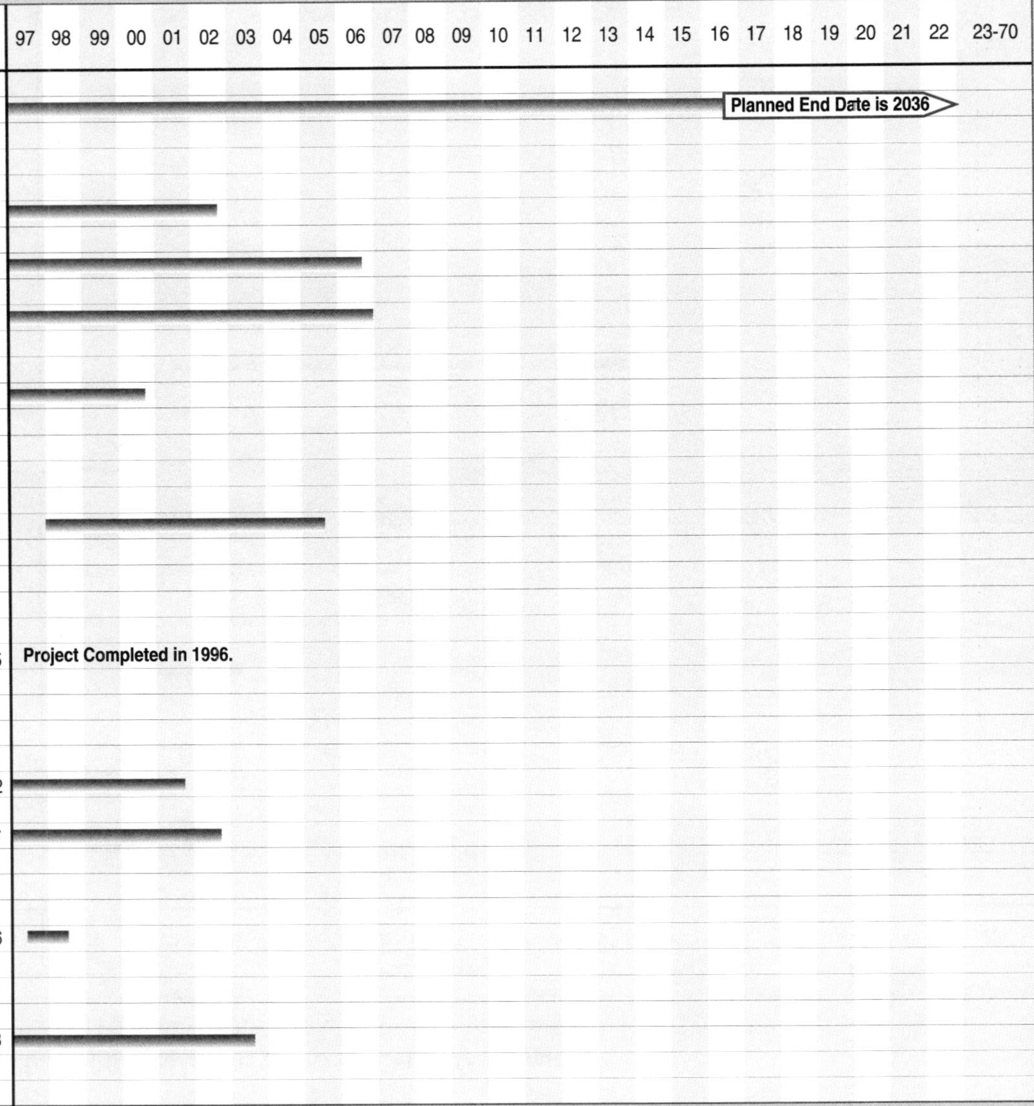

Exhibit E-38 Oakland Operations Office (Continued)
Cleanup Project Summary: Duration and Costs (All costs in thousands of 1998 dollars)

Site Closure Project Activities	1997 - 2006	2007 - 2070	Total
LBNL Legacy Waste	8,240	0	8,240
LBNL Newly Generated Wastes*	17,025	0	17,025
Lawrence Livermore National Laboratory (LLNL)			
LLNL Decontamination and Water Treatment Facility	29,216	0	29,216
Accelerated Waste Treatment	11,911	0	11,911
LLNL Main Site Remediation	144,243	31,902	176,145
LLNL Site 300 Remedial Action	76,531	42,413	118,944
LLNL General Plant Projects*	1,249	0	1,249
LLNL Base Program*	64,521	0	64,521
Separations Process Research Unit (SPRU)			
Separations Process Research Unit	51,605	131,144	182,749
Stanford Linear Accelerator Center (SLAC)			
SLAC Environmental Restoration	5,081	0	5,081
Total	**676,957**	**282,388**	**959,345**

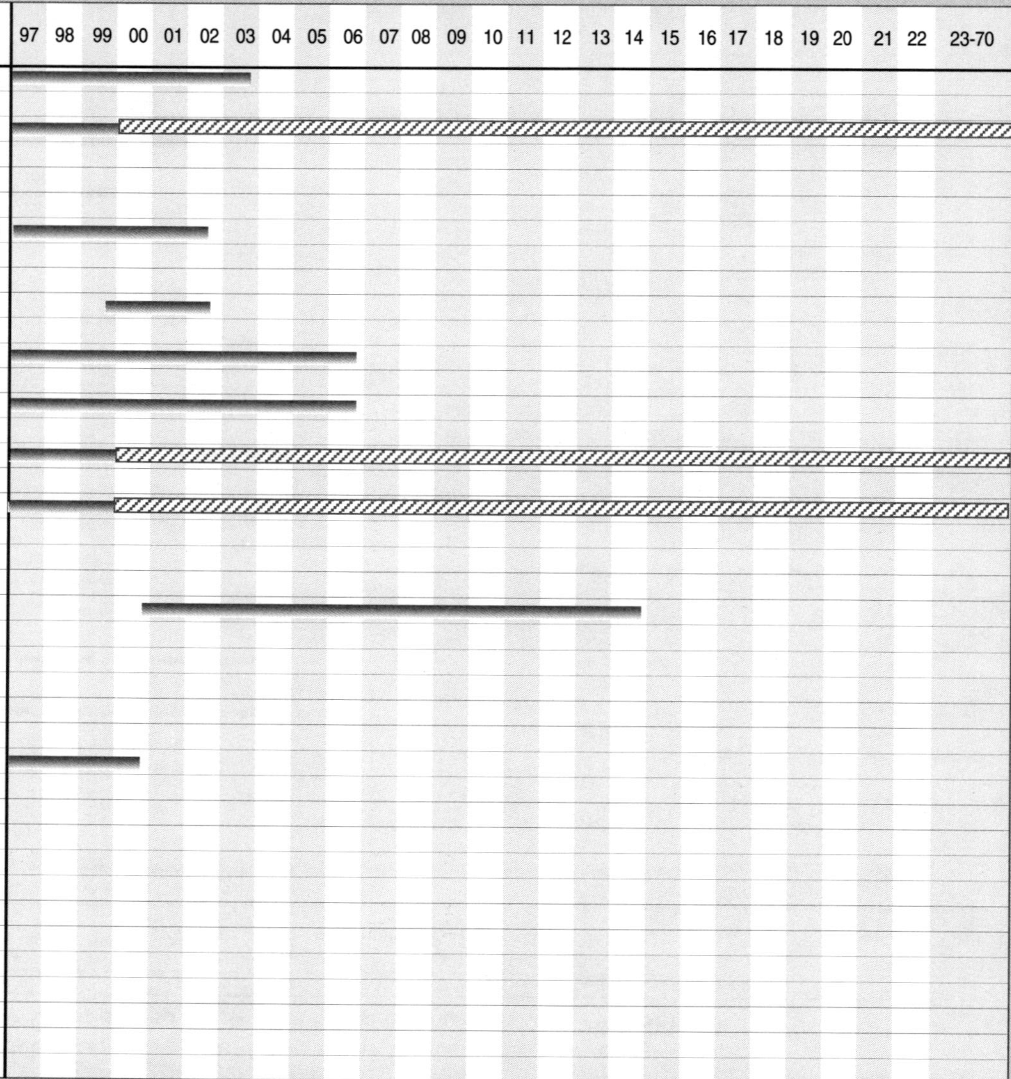

*Project extends through 2070, but newly generated waste costs beginning in FY 2000 are expected to be transfered back to the generator.

The overall planned site completion dates of EM work scope (excluding long-term surveillance and monitoring) are as follows:

Site	Date
Energy Technology Engineering Center	2006
General Atomic Sites	2000
General Electric Vallecitos Nuclear Center	2005
Geothermal Test Facility	1997
Laboratory for Energy-Related Health Research	2002
Lawrence Berkeley National Laboratory	2003
Lawrence Livermore National Laboratory Main Site	2006
Lawrence Livermore National Laboratory Site 300	2006
Separations Process Research Unit	2014
Stanford Linear Accelerator Center	2000

The projected cost profile for environmental management associated with the Oakland Operations Office is developed by combining the cost estimates in each of the PBSs. Exhibit E-39 displays the resultant baseline cost profile.

Exhibit E-39
Oakland Operations Office
Environmental Management Baseline Cost Profile

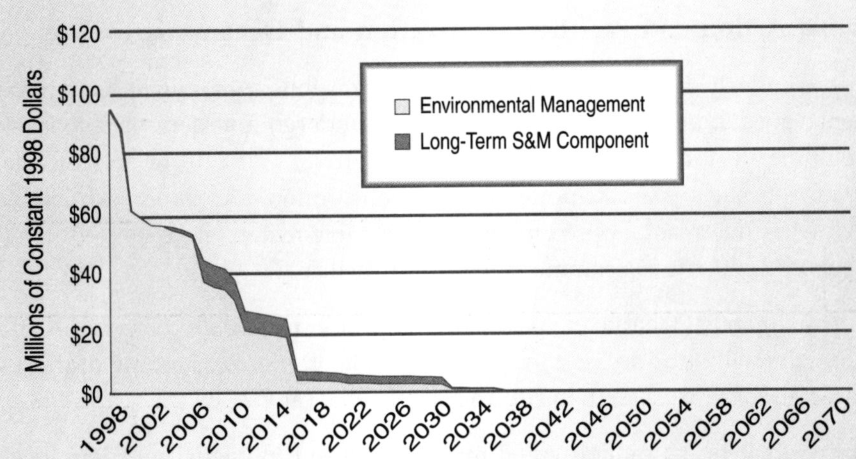

E.7.3 Work Scope Summary

The EM cleanup mission at Oakland Operations Office involves work at nine remaining sites (GTF was completed in FY 1997). Cleanup activities at these sites include the management of groundwater contaminated with volatile organic compounds at Lawrence Livermore National Laboratory and the management of transuranic waste at SPRU. The sections below describe the major waste, material, and contaminated media volumes to be addressed by the Oakland Operations Office. The volumes reported are approximate, and correspond to the major waste, material, and media flows, potential treatment processes, and off-site disposal destinations presented in Exhibit E-40, the Oakland Operations Office Conceptual Summary Disposition Map.

Transuranic Waste

- Approximately 300 cubic meters of legacy transuranic waste are currently in inventory and 880 cubic meters are expected to be generated over the life cycle of operations. After characterization, repackaging, and size reduction, approximately 1,200 cubic meters are expected to be disposed of at WIPP.

Other Waste

- Approximately 470 cubic meters of mixed low-level waste are currently in inventory and 13,000 cubic meters are expected to be generated over the life cycle of operations. After treatment, 8,200 cubic meters are expected to be disposed of at an undetermined facility.

- Approximately 4,200 cubic meters of low-level waste are currently in inventory and 58,000 cubic meters are expected to be generated over the life cycle of operations, of which 660 cubic meters are expected to be reused or recycled. The remainder will be processed, and 60,000 cubic meters are expected to be disposed of off site at either the Nevada Test Site, Hanford, or a commercial disposal facility.

Remedial Action and Facility Deactivation and Decommissioning

- Approximately 43 million cubic meters of hazardous contaminated environmental media, including groundwater, will undergo a variety of responses including in-situ treatment, institutional controls, and on-site and off-site treatments such as air stripping, charcoal absorption, and vapor extraction. Following treatment, approximately 21,000 cubic meters are expected to be disposed of off-site at a commercial disposal facility.

- Approximately 70 cubic meters of transuranic contaminated environmental media will be addressed over the life cycle of operation, some of this is expected to be processed on site and disposed of at WIPP.

- Approximately 2.1 million cubic meters of mixed low-level and low-level contaminated environmental media will be managed and treated. Nearly 8,600 cubic meters are expected to be disposed of off site at a DOE site or a commercial disposal facility.

Exhibit E-40 Oakland Operations Office Conceptual Summary Disposition Map

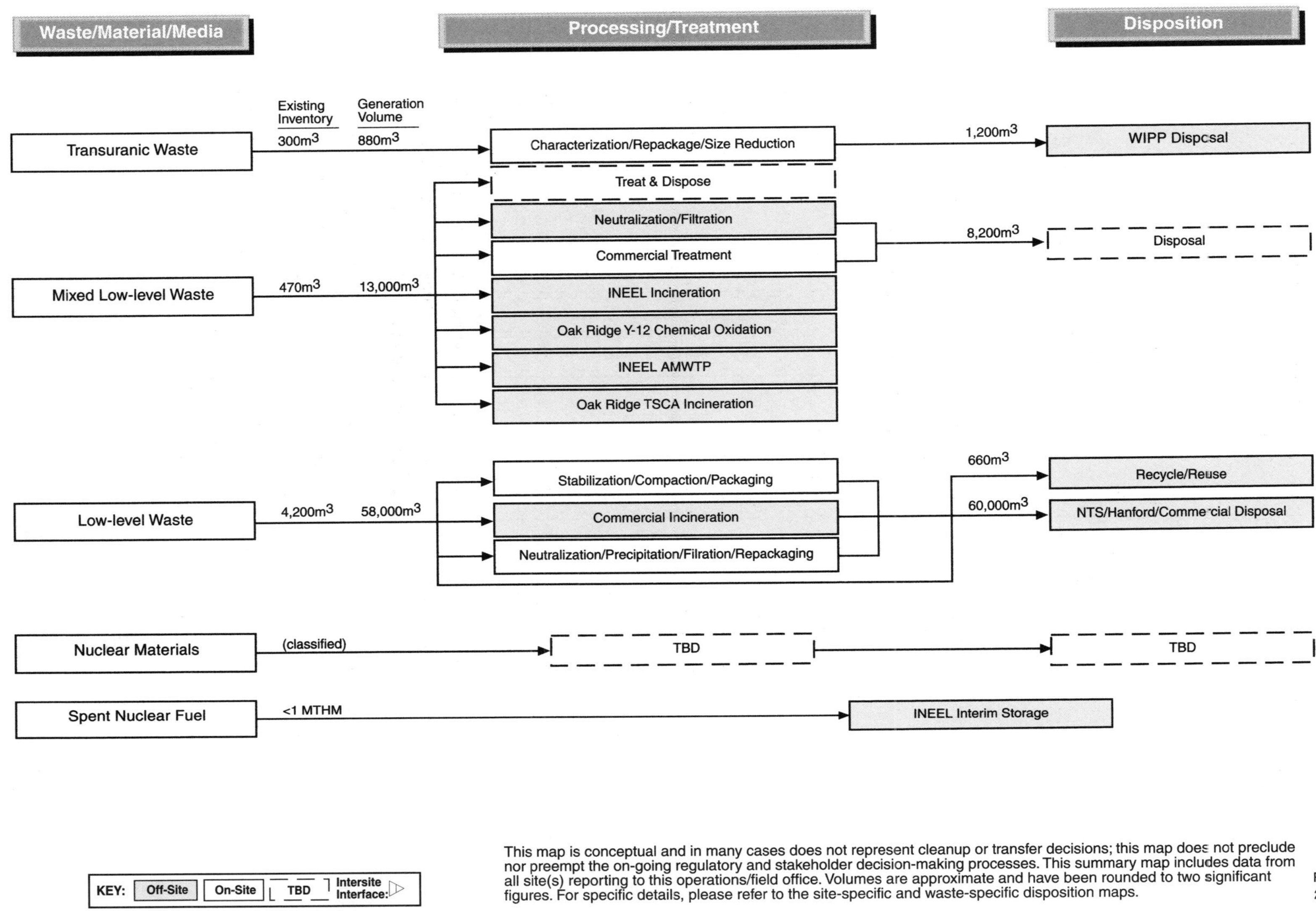

This map is conceptual and in many cases does not represent cleanup or transfer decisions; this map does not preclude nor preempt the on-going regulatory and stakeholder decision-making processes. This summary map includes data from all site(s) reporting to this operations/field office. Volumes are approximate and have been rounded to two significant figures. For specific details, please refer to the site-specific and waste-specific disposition maps.

Rev. 4.0
2/20/98

Exhibit E-40 Oakland Operations Office Conceptual Summary Disposition Map (Continued)

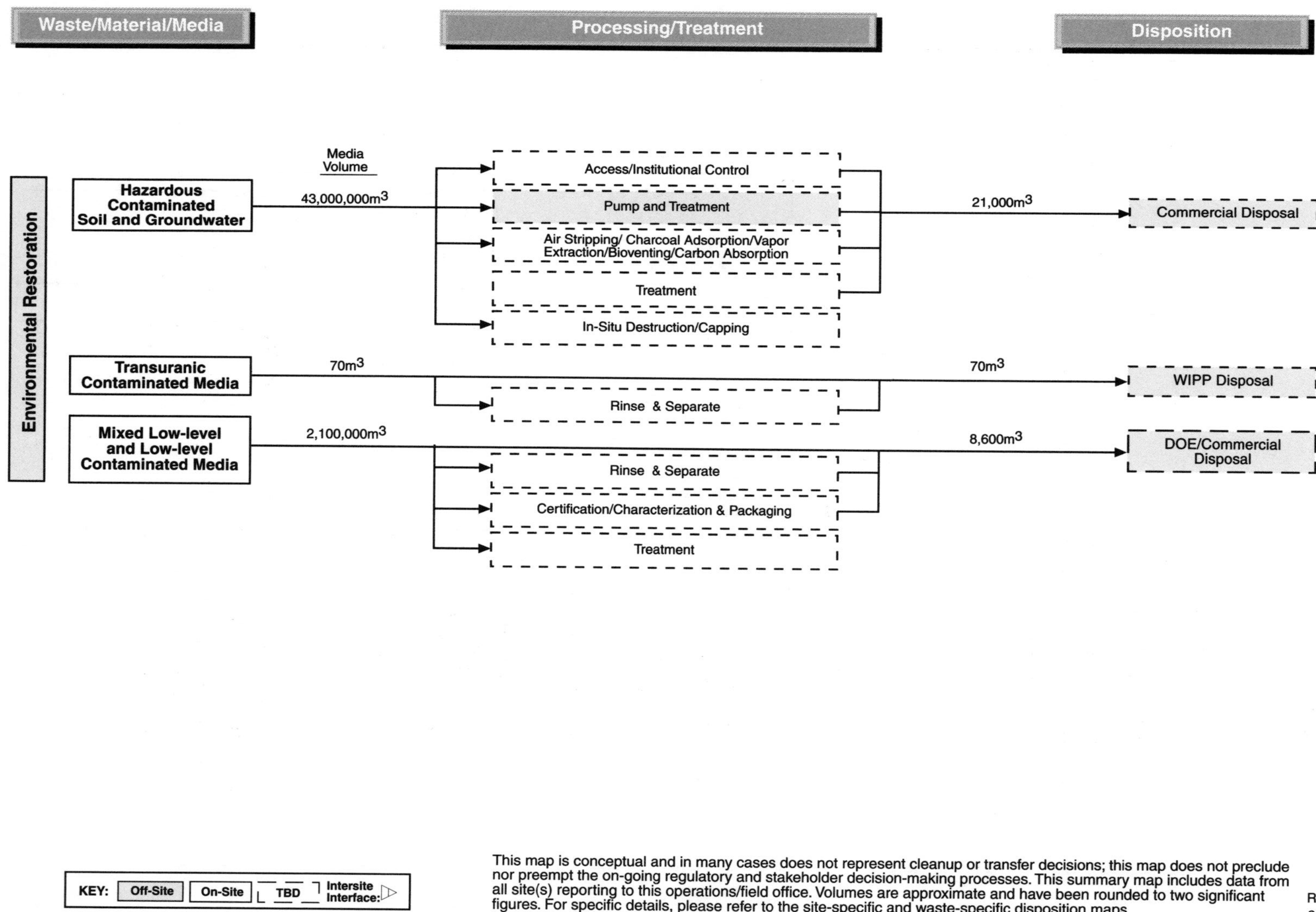

This map is conceptual and in many cases does not represent cleanup or transfer decisions; this map does not preclude nor preempt the on-going regulatory and stakeholder decision-making processes. This summary map includes data from all site(s) reporting to this operations/field office. Volumes are approximate and have been rounded to two significant figures. For specific details, please refer to the site-specific and waste-specific disposition maps.

Rev. 4.0
2/20/98

Nuclear Materials

- Nuclear materials quantities are sensitive and cannot be disclosed in this document.

Spent Nuclear Fuel

- Less than one metric ton heavy metal of spent nuclear fuel will be stabilized and then shipped off site to the Idaho National Engineering and Environmental Laboratory for interim storage.

Exhibit E-41 illustrates Oakland Operations Office environmental management costs by major work scope category. Most costs after 2006 are associated with long-term surveillance and monitoring and the decontamination and decommissioning of SPRU.

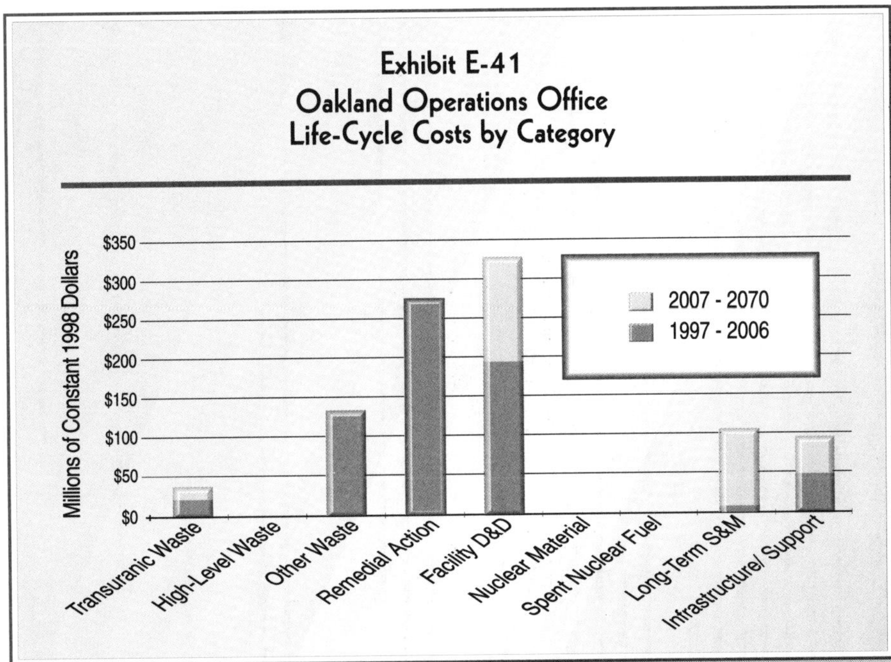

E.7.4 Critical Closure Path and Programmatic Risk

The critical closure path schedule presented as Exhibit E-42 sets forth the timetable for completing the closure activities at Oakland Operations Office, where the bars represent critical activities. The Oakland Operations Office critical closure path reflects those cleanup activities which are key to achieving completion of the site's cleanup mission and end states.

Exhibit E-42

Oakland Operations Office Critical Closure Path

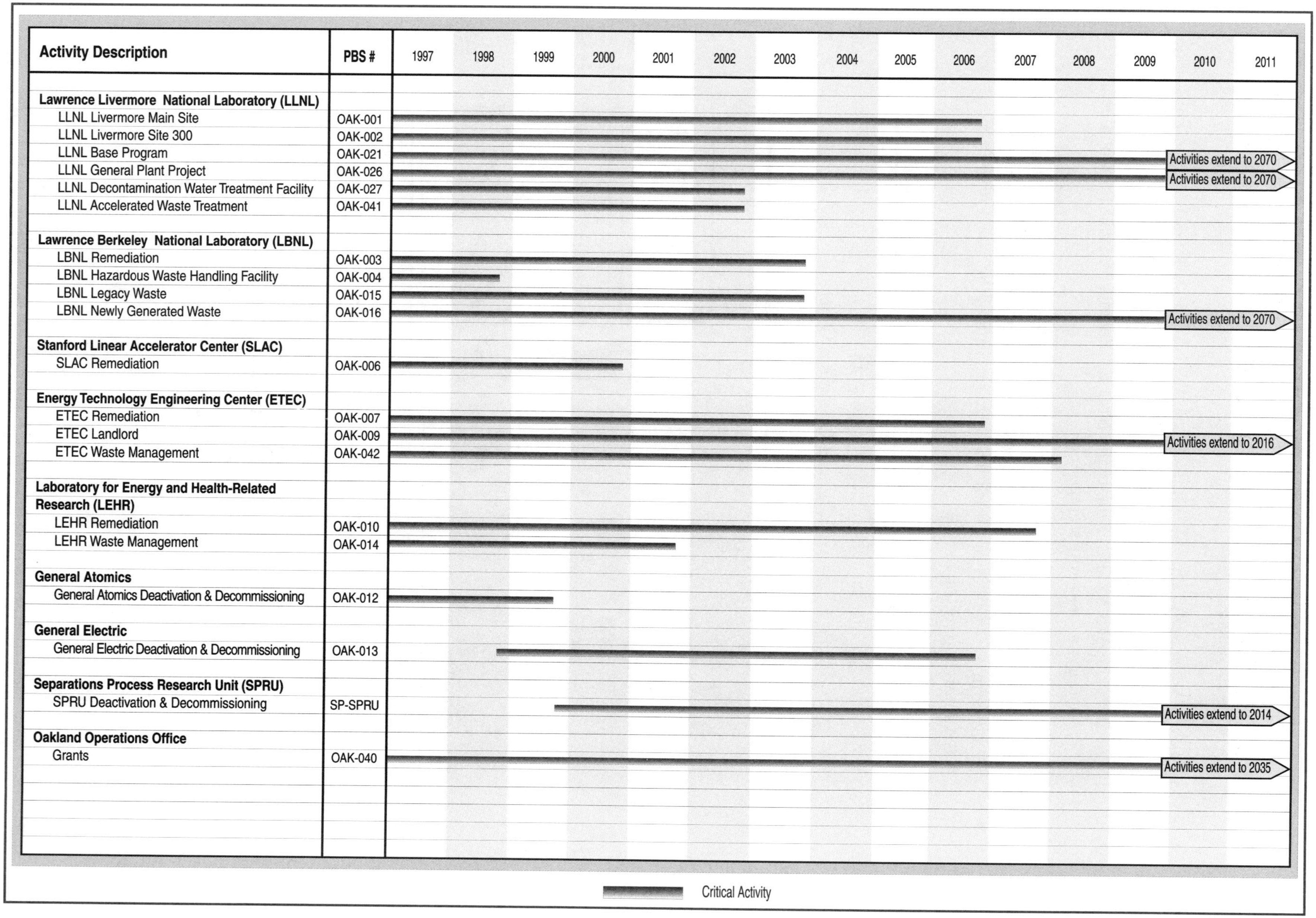

Completion of the EM mission at the Oakland Operations Office as scheduled will depend on the timely accomplishment of critical activities. Appendix D provides a complete definition of programmatic risk. Exhibit E-43 presents a summary of activities/milestones on the critical closure path that have high programmatic risk (programmatic risk scores of 4 or 5 in any category). The Oakland Operations Office version of *Paths to Closure* provides more details on the management approach for these high programmatic risk issues.

Exhibit E-43

Summary of High Programmatic Risk Activities/Milestones: Oakland Operations Office

Site	Project, Activity, Event	Start/End Dates	Programmatic Risk Categories		
			Technological	Work Scope Definition	Intersite Dependency
GA	Package and ship irradiated fuel materials to INEEL for interim storage	Jun 95/ Mar 00	1	4	4
LBNL	Complete characterization and certification for off site disposal of all LBNL legacy waste.	Oct 95/ Sep 70	1	4	3
	Completion of Western Dog Pens Area Removal Action	Apr 00/ Sep 00	2	3	4
LEHR	Completion of disposal of mixed low-level waste/material	May 01/ Aug 01	2	3	4
	No Action ROD for DOE Areas	Sep 01/ Jun 02	NA	4	3

E.8 Ohio Field Office Summary

The Ohio Field Office manages six sites in the states of Ohio and New York. These sites include: Ashtabula Environmental Management Project (RMI Extrusion Plant); Columbus Environmental Management Project (Battelle Columbus Laboratories, two sites); Fernald Environmental Management Project; Miamisburg Environmental Management Project (Mound Plant); and West Valley Demonstration Project (WVDP). Ohio's current baselines reflect completion of its environmental management cleanup mission at all sites by 2008. However, through acceleration and enhanced performance, the goal is to finish by 2005.

Ohio Field Office

The Ohio Field Office manages environmental management activities at six sites in the states of Ohio and New York. These sites include: Ashtabula Environmental Management Project; Columbus Environmental Management Project (2 sites); Fernald Environmental Management Project; Miamisburg Environmental Management Project;

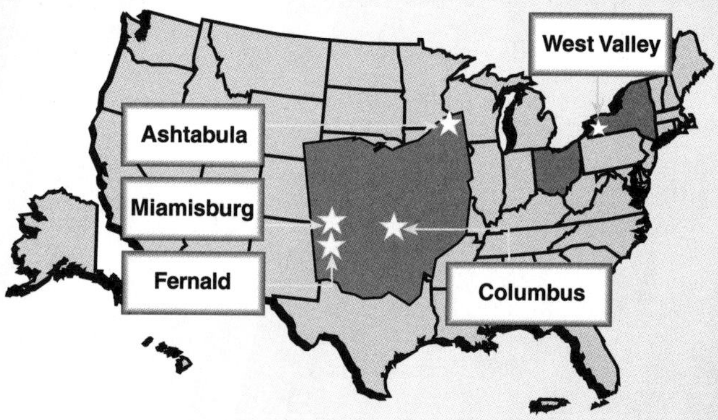

The ***Ashtabula Environmental Management Project*** encompasses the cleanup activities at the RMI Titanium Company Extrusion Plant (formerly Reactive Metals, Inc.), a privately owned facility. From 1962 to 1988, the company received uranium billets and refined them into various shapes for fuel and target fabrication use by the Department of Energy (DOE) and its predecessor agencies. RMI also performed work for the Department of Defense and a number of commercial entities under a Nuclear Regulatory Commission License. Twenty-six years of handling, extruding, forging, and machining uranium at the facility have resulted in on-site and off-site contamination of buildings and environmental media.

The ***Columbus Environmental Management Project*** decommissioning project consists of 15 buildings and includes two geographically distinct sites (West Jefferson and King Avenue). Between 1943 and 1986, Battelle Memorial Institute (Battelle) performed atomic energy research and development for DOE and its predecessor agencies. As part of the Government's fuel and target fabrication program, Battelle participated in nuclear research activities that included fabrication of uranium and fuel elements; reactor development; submarine propulsion; fuel reprocessing; and safety studies of reactor vessels and piping.

The uranium metal production operation at ***Fernald Environmental Management Project*** was constructed in the early 1950s to convert uranium ore into uranium metal, and to fabricate the uranium metal into target elements for reactors that produced weapons-grade plutonium and tritium. Production operations continued for more than 36 years, until DOE suspended them on July 10, 1989.

In 1947, the Dayton Project of the Manhattan Engineering District became the Mound site. Cleanup activities at the Mound site are carried out under the ***Miamisburg Environmental Management Project***. Mound's early mission included nuclear materials research. Later missions included process development, production engineering, manufacturing and surveillance of detonators, explosive timers, transducers, firing sets, explosive pellets, components, and specific test equipment. Additional manufacturing activities at Mound included recovering and purifying tritium.

From 1966 to 1972, Nuclear Fuel Services, Inc., operated a commercial nuclear fuel reprocessing plant at the Western New York Nuclear Services Center under contract to the State of New York. The plant, now referred to as the ***West Valley Demonstration Project***, reprocessed uranium and plutonium from spent nuclear fuel, generating approximately 2.3 million liters (600,000 gallons) of liquid high-level waste that was stored in underground tanks. In 1972, nuclear fuel reprocessing operations were discontinued.

E.8.1 End State

Each of the sites under the Ohio Field Office has a plan in place for end state and long-term stewardship. Exhibit E-44 provides a summary of the anticipated site end states for the Ohio Field Office.

Exhibit E-44

Summary of Ohio Field Office End States

Site Name	End State Description
Columbus Environmental Management Project - King Avenue	King Avenue will be complete in FY 1998, and all 9 buildings and grounds will return to Battelle for reuse without radiological restrictions, according to Nuclear Regulatory Commission (NRC) guidelines. All waste streams, primarily uranium and thorium, will be shipped off site for disposal. The entire Columbus Environmental Management Project will be complete by FY 2005.
Columbus Environmental Management Project - West Jefferson	This site will be complete in FY 2005. The end state will return the buildings and adjacent soil areas at this site back to Battelle in a condition for use without radiological restrictions, according to NRC guidelines. All waste streams will be shipped off site for treatment, storage, or disposal.

Exhibit E-44 (Continued)
Summary of Ohio Field Office End States

Site Name	End State Description
Fernald Environmental Management Project (FEMP)	FEMP will be left in an end state agreed to by the Fernald Citizens Advisory Board and the Community Reuse Organization, although it will still fall under federal ownership. Stakeholders have recommended that specific future use of the site should be determined closer to the time of reuse, but residential and agricultural activities should be avoided. The greatest potential for future use is recreational and industrial. The current FEMP baseline projects that the site will be completed by 2008. However, the Ohio Field Office and the FEMP Office are committed to accomplishing the completion scheduled for 2008 by the end of FY 2005. FEMP will construct a large on-site disposal facility to contain up to 2.5 million cubic yards of low-level wastes with radiological and/or chemical concentrations exceeding free release limits. There will be controlled access to the disposal facility. By 2008, FEMP will install infrastructure to restore the aquifer to a 20 parts per billion (ppb) uranium contamination level through extraction and treatment of groundwater.
Miamisburg Environmental Management Project (MEMP)	Soil remediation to industrial use levels (of approximately 1×10^{-5} reduced risk) will be completed at the Mound Plant in 2003, at which time the site will be sold to the Miamisburg Mound Community Improvement Corporation (MMCIC). The Miamisburg Mound Community Improvement Corporation was formed in order to effectively represent the interests of the local community. Environmental Management will remain the landlord, though the Office of Nuclear Energy (NE) will have a continuing mission at Mound through its use of seven buildings. The landlord costs and cleanup requirements for these buildings are the responsibility of NE. Volatile organic compound-contaminated off-site groundwater will be remediated to a residential level prior to FY 2005. Excess nuclear materials will be off site in FY 1998. Currently, MEMP is planned for completion by 2005. Pending validation of the current baseline, it is the goal of the Ohio Field Office and the MEMP Office to clean up the site in 2003.
Ashtabula Environmental Management Project (AEMP)	The end state for the AEMP will be reached in 2003 when the site will be released to RMI. RMI will have sole responsibility for future land use. Future use is assumed to be industrial, consistent with surrounding property and zoning. Surficial soils contaminated with uranium will be remediated to less than 30 pCi/g. The NRC license will be terminated in 2003 when the property is released.
West Valley Demonstration Project (WVDP)	The site is owned by New York State but DOE has exclusive use and possession of the WVDP premises. By the end of FY 2005, DOE will have satisfied its responsibilities for West Valley according to the West Valley Demonstration Project Act, Stipulation of Compromise Settlement, the Cooperative Agreement, and the Record of Decision, after which DOE will not be responsible for any of the decisions involving the future use of the site. The end state for the WVDP involves completion of HLW solidification, and shipment of HLW canisters, LLW, MLLW and TRU in accordance with the WVDP Act Stipulation of Compromise and ROD. The SNF will be shipped to INEEL. Tanks and facilities will be decontaminated and decommissioned. Operational responsibility will be returned to the New York State Energy Research and Development Authority (NYSERDA). LLW disposal has yet to be determined.

E.8.2 Cost and Completion Dates

Ohio Field Office has divided its environmental management work into 31 discrete projects. A Project Baseline Summary (PBS) exists for each project and contains detailed programmatic information, including cost, schedule, scope, end state, and interim milestones. A summary of the Ohio cost and schedule information is illustrated in Exhibit E-45. For additional information on these projects, refer to the individual PBSs.

The estimated Office of Environmental Management (EM) life-cycle cost of Ohio Field Office site cleanups is $4.8 billion (constant 1998 dollars) with the last project ending in 2008. Groundwater remediation and some surveillance and monitoring will continue beyond 2008 at some sites.

The overall site planned completion dates are as follows:

Site	**Date**
Columbus Environmental Management Project West Jefferson Site	2005
Columbus Environmental Management Project King Avenue Site	1998
Fernald Environmental Management Project	2008
Miamisburg Environmental Management Project	2005
Ashtabula Environmental Management Project	2003
West Valley Demonstration Project	2005

The projected cost profile for environmental management associated with the Ohio Field Office is developed by combining the cost estimates in each of the Project Baseline Summaries. Exhibit E-46 displays the resultant baseline cost profile.

E.8.3 Work Scope Summary

EM's mission at Ohio consists of various projects focused on the general tasks of decontamination, deactivation, excavation and treatment of contaminated soils, groundwater remediation, the vitrification of high-level waste (West Valley), along with many others. At the Columbus Environmental Management Project King Avenue site, the major work scope revolves around the decontamination of the remaining buildings. The decontamination approach for the buildings follows a standard flow beginning with a physical and radiological survey and ending with the full completion of the decontamination after proceeding through a series of prescribed steps. At the Columbus Environmental Management Project West Jefferson site, a significant effort will be required to process highly contaminated equipment and materials prior to beginning interior decontamination. However, there are a few facilities at the West Jefferson site, the JN-1 hot cells, which will involve a more extensive effort, using remote-controlled operations to reduce levels of contamination within highly

Exhibit E-45 Ohio Field Office
Cleanup Project Summary: Duration and Costs (All costs in thousands of 1998 dollars)

Site Closure Project Activities	1997 - 2006	2007 - 2070	Total
Ashtabula Environmental Management Project			
Project Management, Site Services, Environmental Safety and Health	28,247	0	28,247
Ashtabula Remediation	64,335	642	64,977
Columbus Environmental Management Project			
King Avenue Site Decontamination	17,955	0	17,955
West Jefferson Site Decontamination	93,587	0	93,587
Project Management, Site Support & Maintenance	27,453	0	27,453
Fernald Environmental Management Project			
Thorium Overpack	1,695	0	1,695
Facility Shutdown	273,262	5,377	278,639
Nuclear Materials	5,879	0	5,879
Mixed Waste	21,843	0	21,843
Waste Management	87,578	3,572	91,150
Waste Pits Remediation Project	373,961	0	373,961
Facility Deactivation and Decommissioning	178,216	0	178,216
Soils	182,021	0	182,021
On-site Disposal Facility	187,436	1,553	188,989

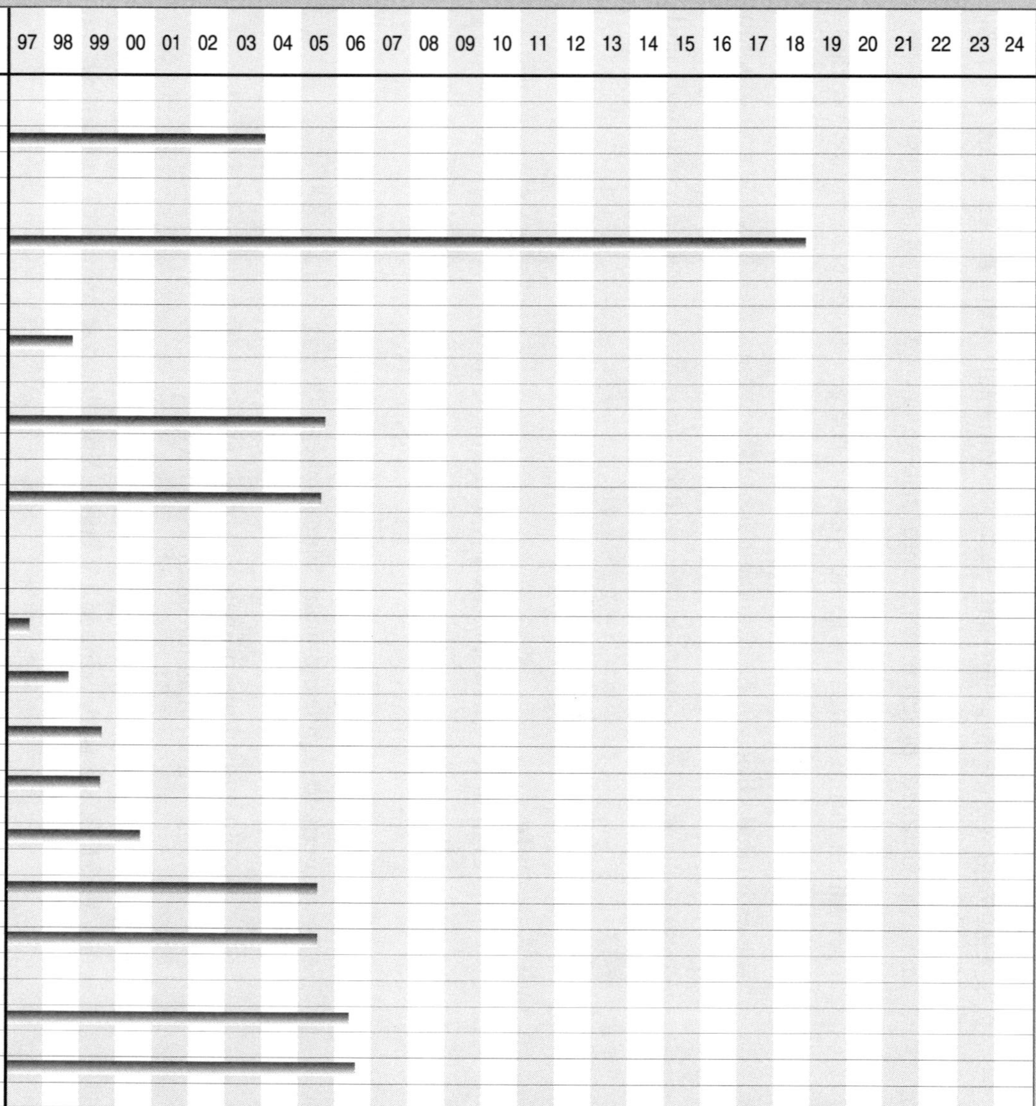

Exhibit E-45 Ohio Field Office (Continued)
Cleanup Project Summary: Duration and Costs (All costs in thousands of 1998 dollars)

Site Closure Project Activities	1997 - 2006	2007 - 2070	Total
Aquifer Restoration	220,359	22,828	243,187
Program Support & Oversight	651,687	66,225	717,912
Silos	331,440	74,498	405,938
Miamisburg Environmental Management Project			
Tritium Operations Transition	20,198	0	20,198
Main Hill Tritium	55,447	0	55,447
Legacy Waste	17,011	0	17,011
Main Hill Rad	21,016	0	21,016
Special Materials/ Plutonium Processing Hill	22,103	0	22,103
Soils	44,610	571	45,181
Main Hill Non Rad	8,998	0	8,998
Test Fire Valley	27,800	0	27,800
Facility Operations & Maintenance	64,628	0	64,628
Exit Support Project	516,404	0	516,404
West Valley Demonstration Project			
High-Level Waste Vitrification and Tank Heel High Activity Waste Processing	295,899	0	295,899
Site Transition, Decommissioning, & Project Completion	482,882	0	482,882
Spent Nuclear Fuel	20,692	0	20,692
Project Management/Site Support	314,213	0	314,213
Total	**4,658,852**	**175,266**	**4,834,118**

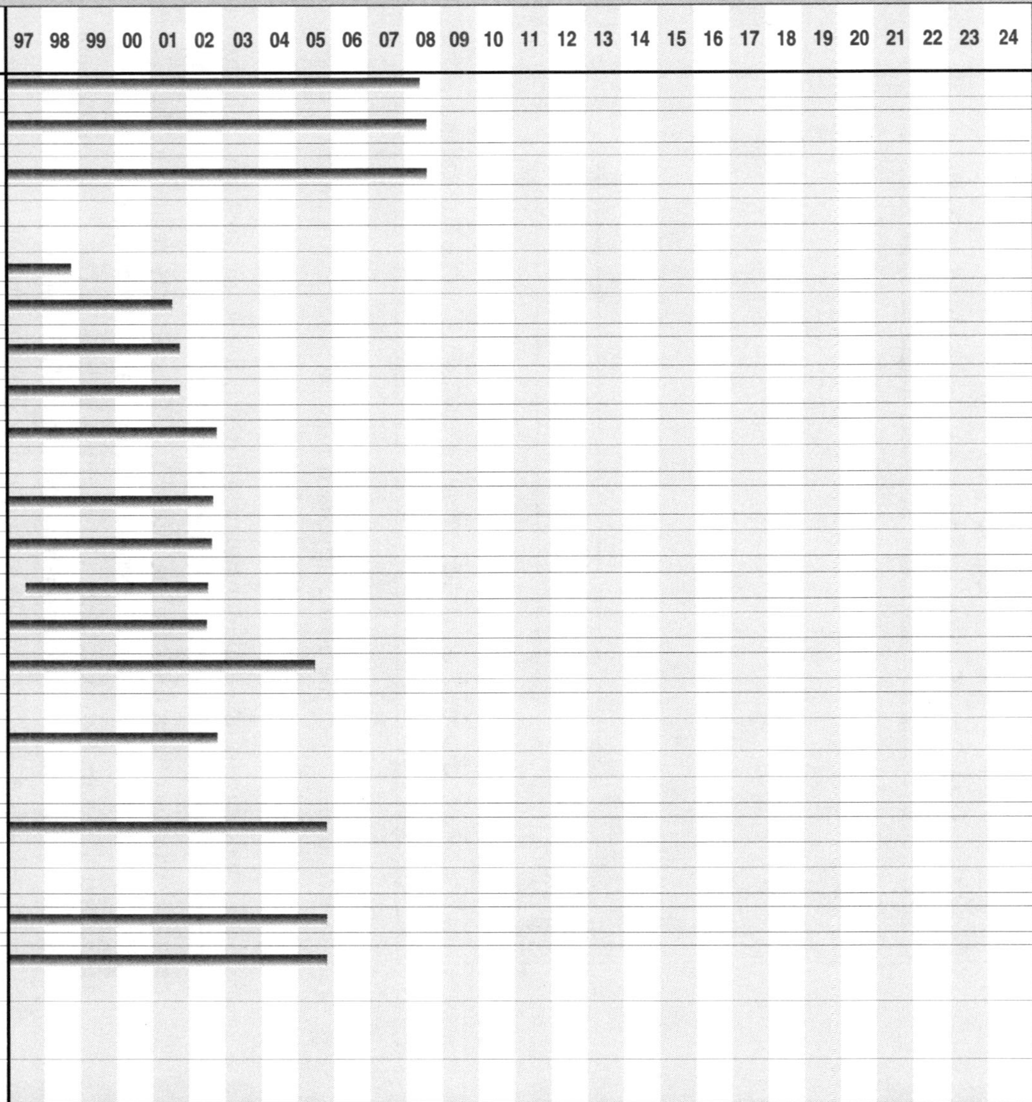

Exhibit E-46
Ohio Field Office Environmental Management Baseline Cost Profile

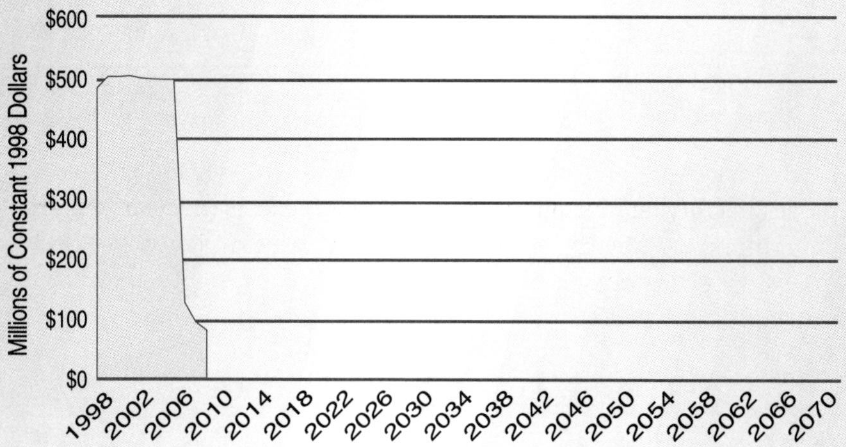

radioactive areas. Also, the actual approach may be modified depending on the end use planned for the West Jefferson buildings.

At the Fernald Environmental Management Project, the scope, cost, and schedule reflected in *Paths to Closure* are as documented in the project baseline. The principal work scope in the baseline after FY 2005 is directly related to the Silos Project, Facilities Shutdown, Decontamination and Decommissioning, and associated Program Support and Oversight activities. The most significant challenge Fernald faces in accomplishing the Ohio 2005 Vision is accelerating the Silos Project. Once the Fernald Environmental Management Project is completed, the only remaining activities include closure of the On-Site Disposal Facility, finalization of waste management activities and closure of facilities, and in-process groundwater monitoring.

At the Ashtabula Environmental Management Project, the remediation work scope of the RMI Extrusion facility will involve the deactivation of 25 on-site buildings and decontamination and/or demolition of 21; remediation of legacy waste and associated equipment; excavation and treatment/processing of radiologically contaminated soils; and ex-situ vapor stripping of groundwater.

At the West Valley Demonstration Project, the baseline consists of four projects. The first project encompasses the work scope involved in the solidification of high- level waste into borosilicate glass using vitrification. Following this, the WVDP plans to process the tank residual high activity waste. The second project encompasses activities required for removal of high-level waste canisters and transuranic waste from project facilities, disposal of low-level waste and mixed low-level waste in accordance with the Act and Stipulation of Compromise as directed by the final Environmental Impact Statement Record of Decision, and disposition of the remaining project responsibilities. The third project encompasses the work scope involved with the removal of the existing spent nuclear fuel inventory from the site. The fourth project encompasses the general mission and support cost estimates relating to project management, human resources, program planning, Chief Financial Officer, procurement, financial control, information services, training, records management, legal and program reporting functions. These four projects make up the work scope for the West Valley Demonstration Project.

At the Miamisburg Environmental Management Project, the work scope encompasses facility stabilization, disposition of excess nuclear material and ancillary equipment, environmental restoration, decommissioning, and waste management. The disposition of nuclear materials, including tritium, is targeted for completion in FY 1998.

The sections below describe the major waste, material, and contaminated media volumes to be addressed by the Ohio Field Office. The volumes reported are approximate, and correspond to the major waste, material, and media flows, potential treatment processes, and off-site disposal destinations presented in Exhibit E-47, the Ohio Field Office Conceptual Summary Disposition Map.

Transuranic Waste

- Approximately 770 cubic meters of transuranic waste are currently in inventory and 24 cubic meters are expected to be generated over the life cycle of operations. After characterization, compaction, and packaging, 250 cubic meters are expected to be disposed of at the Waste Isolation Pilot Plant (WIPP), and a remaining 550 cubic meters are expected to be disposed of at a currently undetermined facility.

High-level Waste

- Approximately 2,200 cubic meters of high-level waste currently in inventory, will be washed and vitrified. After vitrification, 250 cubic meters are expected to be disposed of at a geologic repository.

Other Waste

- Approximately 220 cubic meters of mixed low-level waste are currently in inventory and 38 cubic meters are expected to be generated over the life cycle of operations. After treatment, 9.3 cubic meters are expected to be disposed of at an off-site commercial facility, 1.8 cubic meters are expected to be disposed of at a waste water disposal facility, and 45 cubic meters are expected to be disposed of at an undetermined facility.

Exhibit E-47 Ohio Field Office Conceptual Summary Disposition Map

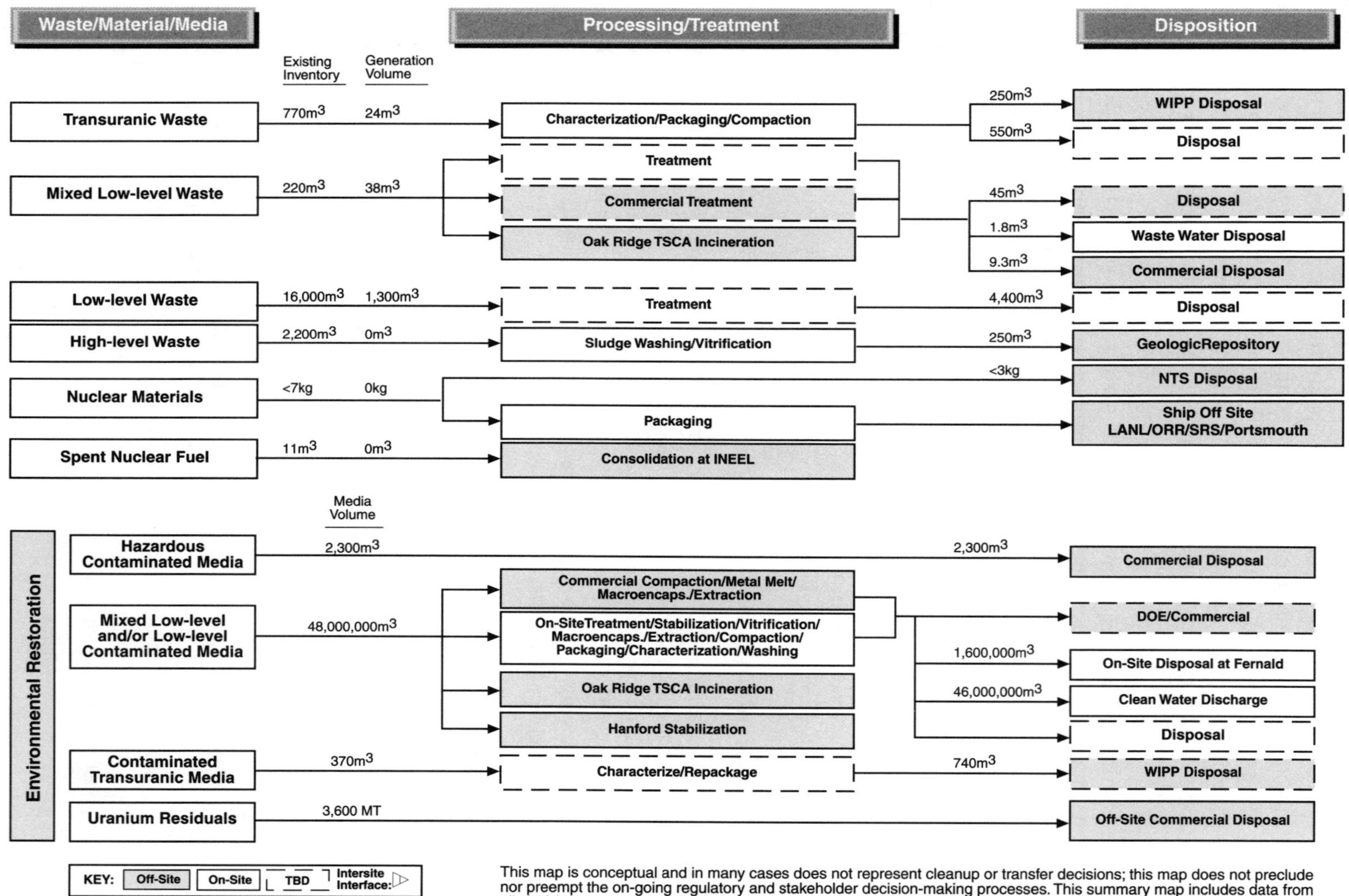

This map is conceptual and in many cases does not represent cleanup or transfer decisions; this map does not preclude nor preempt the on-going regulatory and stakeholder decision-making processes. This summary map includes data from all site(s) reporting to this operations/field office. Volumes are approximate and have been rounded to two significant figures. For specific details, please refer to the site-specific and waste-specific disposition maps.

Rev. 4.0
2/20/98

- Approximately 16,000 cubic meters of low-level waste are currently in inventory and 1,300 cubic meters are expected to be generated over the life cycle of operations. After treatment, 4,400 cubic meters are expected to be disposed of at an undetermined facility.

Remedial Action and Facility D&D

- There are approximately 2,300 cubic meters of hazardous contaminated environmental media which will be disposed of at an off-site commercial disposal facility.
- Approximately 48 million cubic meters of mixed low-level and low-level contaminated environmental media, including groundwater, will go through treatment, incineration, and/or stabilization. Approximately 1.6 million cubic meters of waste are expected to be disposed of on site at Fernald, and approximately 46 million cubic meters of treated water will be discharged. Additional volumes of waste are expected to be disposed of at a DOE site, a commercial facility, or an undetermined location.
- Approximately 370 cubic meters of environmental media contaminated with transuranic elements will be characterized and repackaged, and 740 cubic meters are expected to be disposed of at WIPP.
- Approximately 3,600 metric tons of uranium residuals are expected to be disposed of at an off-site commercial disposal facility.

Nuclear Materials

- Currently, there are less than 7 kilograms of nuclear materials in inventory. Of this amount, less than 3 kilograms will be shipped to the Nevada Test Site for disposal and, after packaging, the remaining amount will be shipped off site to Los Alamos National Laboratory, Oak Ridge Reservation, Savannah River Site, and Portsmouth.

Spent Nuclear Fuel

- Currently, there are 11 cubic meters of spent nuclear fuel in inventory. This waste stream will be shipped off site for consolidation at a commercial disposal facility.

Exhibit E-48 shows the distribution of Ohio Field Office EM costs by major category.

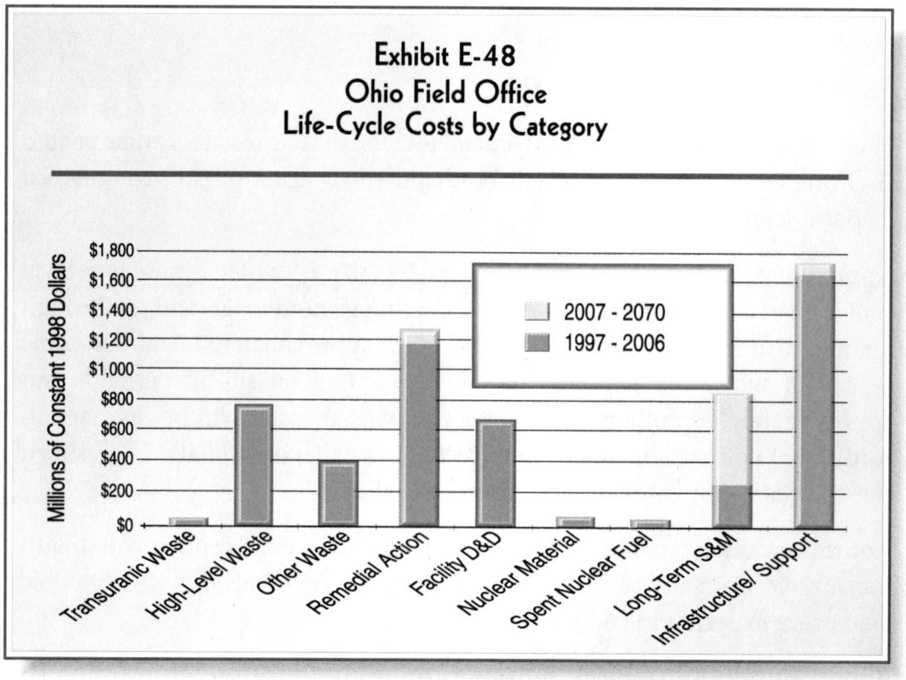

E.8.4 Critical Closure Path and Programmatic Risk

The critical closure path schedule presented as Exhibit E-49 sets forth the timetable for completing the closure activities at the Ohio Field Office. The highlighted activities show the critical closure path, which represents the series of events that drive the overall completion date for the site. In Exhibit E-49, the bars represent critical activities, and the diamonds represent milestones/events.

Completion of the EM mission at the Ohio Field Office as scheduled will depend on stable funding and the timely accomplishment of critical activities and milestones. Sites have assigned programmatic risk scores to each of the critical activities/milestones. Appendix D provides a complete definition of programmatic risk. Exhibit E-50 presents a summary of activities/milestones on the critical closure path that have high programmatic risk (programmatic risk scores of 4 or 5 in any category). The Ohio Field Office version of *Paths to Closure* provides more details on the management approach for these high programmatic risk issues.

Exhibit E-49 Ohio Field Office Critical Closure Path (Timelines and Risk Data)

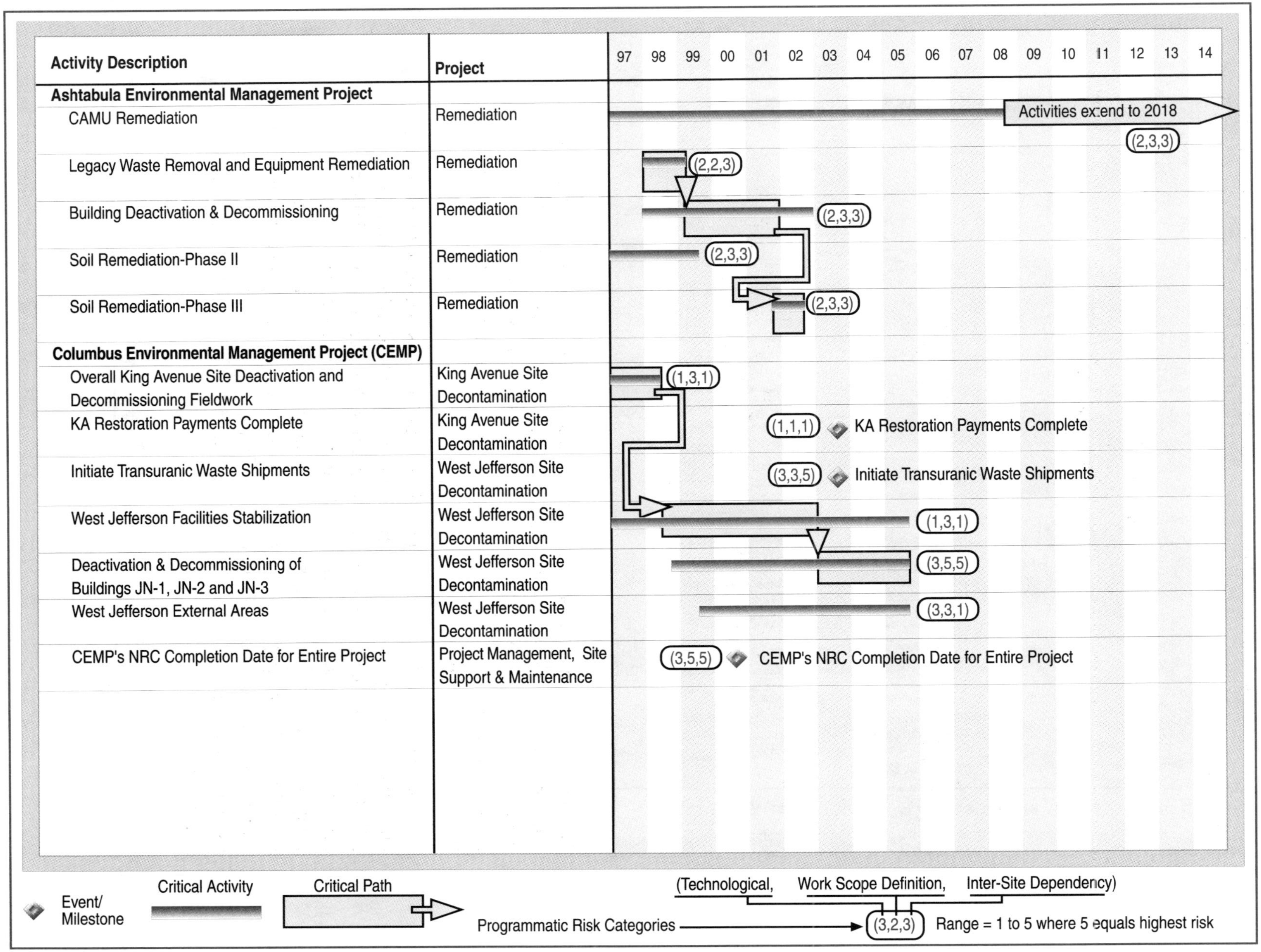

Exhibit E-49 — **Ohio Field Office Critical Closure Path (Continued)** (Timelines and Risk Data)

Activity Description	Project
Fernald Environmental Management Project	
Packaging, Shipping and Disposition of all Nuclear Materials Presently On Site	Nuclear Materials
Characterization, Processing, Packaging, Shipping and Disposition of all Legacy Mixed Waste	Mixed Waste
Remove Nuclear Material from Facilities; Deactivation and Decommissioning Site Facilities	Facility Shutdown, Facility D&D
Excavation and Disposition of Waste Buried in On-site Waste Pits	Waste Pits Remediation Project
Liner Construction, Waste Placement (Soil and Debris), and OSDF Capped	On-site Disposal Facility
Processing and Disposition of the Waste Contained in the K-65 Silos	Silos
Miamisburg Environmental Management Project	
Complete Shipment of Bulk Accountable Tritium	Tritium Operations Transition
Decontaminate T Building. Safe Shutdown of SW and R.	Main Hill Tritium
Removal of TRU Waste from T Building and Ship Off Site	Legacy Waste
Complete Building R and SW Decommissioning	Main Hill Rad
Complete E/E Annex Decommissioning	Main Hill Rad
Complete D&D/Closure of Waste Processing Facilities (19, 72, 57, and 22)	Test Fire Valley
West Valley Demonstration Project (WVDP)	
Select High-level Waste Receiving Site	Site Transition, Decommissioning, & Project Completion
DOE-HQ to Approve Supplement to Draft Environmental Impact Statement (EIS)	Site Transition, Decommissioning, & Project Completion
DOE-HQ to Issue WVDP Completion Record of Decision (ROD)	Site Transition, Decommissioning, & Project Completion

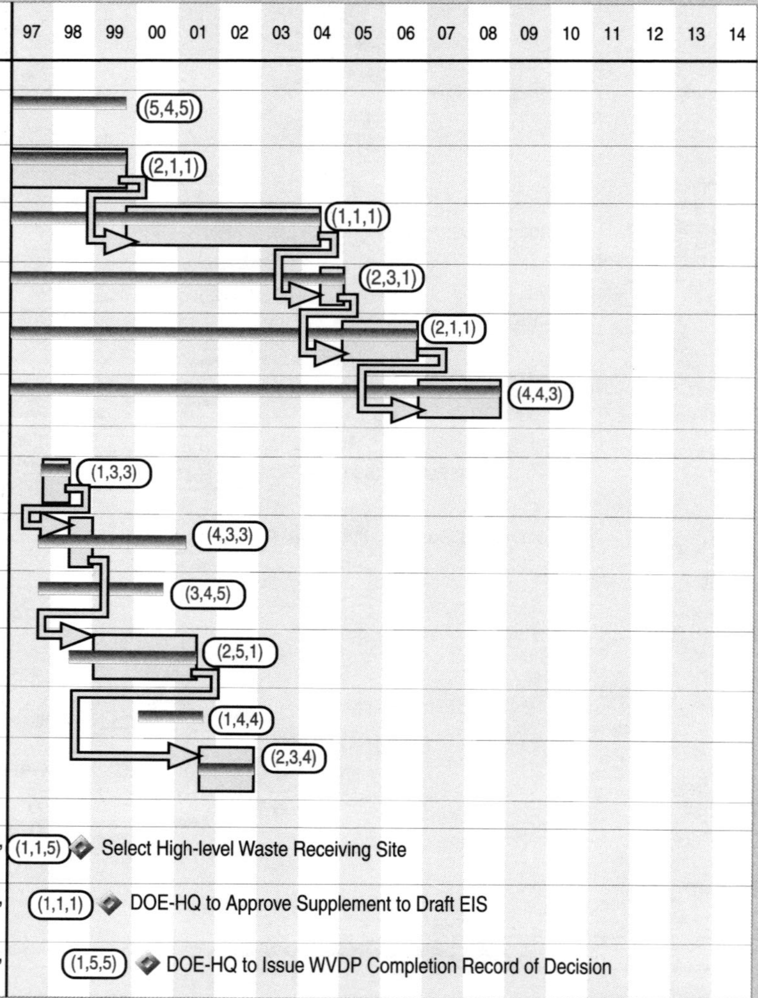

Event/ Milestone — Critical Activity — Critical Path

Programmatic Risk Categories → (3,2,3) (Technological, Work Scope Definition, Inter-Site Dependency) Range = 1 to 5 where 5 equals highest risk

Exhibit E-49 **Ohio Field Office Critical Closure Path (Continued)** (Timelines and Risk Data)

Activity Description	Project	97–14 (Risk Data)
DOE-HQ to Approve Final EIS	Site Transition, Decommissioning, & Project Completion	(1,1,1) DOE-HQ to Approve Final EIS
Select Transuranic Waste Receiving Site	Site Transition, Decommissioning, & Project Completion	(2,5,5) Select Transuranic Waste Receiving Site
Spent Nuclear Fuel Prerequisite Milestones at Idaho National Engineering and Environmental Laboratory (INEEL) Performed	Spent Nuclear Fuel	(2,1,2) Spent Nuclear Fuel Prerequisite Milestones at INEEL Performed
NRC Approval of D&D Criteria and Site Decommissioning Plan	Site Transition, Decommissioning, & Project Completion	(1,4,2) NRC Approval of D&D Criteria and Site Decommissioning Plan
Vitrification Facility/Tank Farm Deactivation Campaign and Phase I&II	HLW Vitrification and Tank Heel High Activity Waste Processing	(2,3,1)
Develop Implementation Plan for Project Completion and Issue ROD	Site Transition, Decommissioning, & Project Completion	(2,5,5)
Ship West Valley Spent Nuclear Fuel to INEEL	Spent Nuclear Fuel	(1,1,1)
Removal of High-level & Transuranic Waste from WVDP	Site Transition, Decommissioning, & Project Completion	(2,5,5)
DOE-HQ Provides TRU & HLW Casks, Permits, & Transition Program Funds	Site Transition, Decommissioning, & Project Completion	(2,4,4)
Begin Removal of West Valley Transuranic Waste	Site Transition, Decommissioning, & Project Completion	(2,3,2)
Begin Removal of WVDP HLW Canisters from HLW Interim Storage Facility	Site Transition, Decommissioning, & Project Completion	(2,3,2)
Begin Site Decontamination per NRC Criteria	Site Transition, Decommissioning, & Project Completion	(2,4,2)
Begin Disposition of Vitrification Facility/Tank Farm per NRC Criteria	Site Transition, Decommissioning, & Project Completion	(2,4,2)
Complete Site Decontamination per NRC Criteria	Site Transition, Decommissioning, & Project Completion	(2,5,5)

Event/Milestone — Critical Activity — Critical Path

Programmatic Risk Categories (Technological, Work Scope Definition, Inter-Site Depencency) (3,2,3) Range = 1 to 5 where 5 equals highest risk

Exhibit E-50
Summary of High Programmatic Risk Activities/Milestones: Ohio Field Office

Site	Project, Activity, Event	Start/End Dates	Programmatic Risk Categories		
			Technological	Work Scope Definition	Intersite Dependency
CEMP	NRC completion date for entire project	Sep 00/ Sep 00	3	5	5
	Initiate TRU waste shipments	Oct 03/ Oct 03	3	3	5
	Building JN-1 D&D	Oct 98/ Sep 05	3	5	5
FEMP	Scope includes packaging, shipping and disposition of all nuclear materials presently on site. Scope is not fully funded.	Oct 96/ Sep 99	5	4	5
	Scope includes processing and disposition of the waste contained in the K-65 silos.	Oct 96/ Sep 08	4	4	3
MEMP	Removal of TRU waste from T Building and ship off site	Oct 97/ Oct 00	3	4	5
	Decontaminate T Building. Safe shutdown of SW and R Buildings.	Oct 97/ Apr 01	4	3	3
	Complete Building SW decommissioning	Jun 98/ Jun 01	2	5	1
	Complete Building R decommissioning	Feb 99/ Jul 01	2	5	1
	Complete E/E Annex decommissioning	Feb 00/ Oct 01	1	4	1
	Complete D&D/closure of waste processing facilities (19, 72, 57, and 22)	Oct 01/ Dec 02	2	3	4

Exhibit E-50 (Continued)

Site	Project, Activity, Event	Start/End Dates	Programmatic Risk Categories		
			Technological	Work Scope Definition	Intersite Dependency
WVDP	Select HLW receiving site	Sep 98/ Sep 98	1	1	5
	DOE-HQ to issue WV Project Completion ROD	May 00	1	5	5
	Select TRU receiving site	Jun 00/ Jun 00	2	5	5
	DOE-HQ provides HLW casks, permits, agreements, & transportation program funds	Oct 98/ Mar 01	2	4	4
	Issue WV Project Completion ROD/Develop Implementation Plan for project completion	May 00/ Mar 01	2	5	5
	NRC approval of D&D criteria and Site Decommissioning Plan	Sep 01	1	4	2
	DOE-HQ provides TRU casks, permits, agreements, & transportation program funds	Jul 00/ Sep 02	2	4	4
	Removal of WV HLW & TRU from WV	Oct 01/ Jun 04	2	5	5
	Begin disposition of vitrification facility/tank farm per NRC criteria	Oct 02/ Sep 05	2	4	2
	Begin site decontamination per NRC criteria	Oct 01/ Sep 05	2	4	2
	Complete site decontamination per NRC criteria	Jul 01/ Sep 05	2	5	5

Appendix F

List of References

List of References

U.S. Department of Energy (DOE), December 1997. ***Preliminary Responses to Comments on the Accelerating Cleanup: Focus on 2006 National Discussion Draft*** (DOE/EM-0338). Office of Environmental Management, Washington, DC.

U.S. Department of Energy (DOE), June 1997. ***Accelerating Cleanup: Focus On 2006 Discussion Draft*** (DOE/EM-0327). Office of Environmental Management, Washington, DC.

U.S. Department of Energy (DOE), January 1997. ***Linking Legacies: Connecting the Cold War Nuclear Weapons Production Processes To Their Environmental Consequences*** (DOE/EM-0319). Office of Environmental Management, Washington, DC.

U.S. Department of Energy (DOE), June 1996. ***The 1996 Baseline Environmental Management Report*** (DOE/EM-0290). Office of Environmental Management, Washington, DC.

U.S. Department of Energy (DOE), January 1996. ***Taking Stock: A Look at the Opportunities and Challenges Posed by Inventories from the Cold War Era*** (DOE/EM-0275). Office of Environmental Management, Washington, DC.

U.S. Department of Energy (DOE), March 1995. ***Estimating the Cold War Mortgage: The 1995 Baseline Environmental Management Report*** (DOE/EM-0232). Office of Environmental Management, Washington, DC.

U.S. Department of Energy (DOE), January 1995. ***Closing the Circle on the Splitting of the Atom: The Environmental Legacy of Nuclear Weapons Production in the United States and What the Department of Energy is Doing About It.*** Office of Environmental Management, Washington, DC.

Notes

Notes

ISBN 0-16-049651-9

90000

9 780160 496516

For sale by the U.S. Government Printing Office
Superintendent of Documents, Mail Stop: SSOP, Washington, DC 20402-9328
ISBN 0-16-049651-9